Sexing the Brain

MAPS OF THE MIND

STEVEN ROSE, GENERAL EDITOR

MAPS OF THE MIND

STEVEN ROSE, GENERAL EDITOR

—

Pain: The Science of Suffering
Patrick Wall

The Making of Intelligence
Ken Richardson

How Brains Make Up Their Minds
Walter J. Freeman

Sexing
the Brain

Lesley Rogers

Columbia University Press

New York

Columbia University Press
Publishers Since 1893
New York   Chichester, West Sussex

Copyright © 2001 Lesley Rogers
First published by Weidenfeld and Nicolson Ltd., London

Library of Congress Cataloging-in-Publication Data
Rogers, Lesley.
Sexing the brain / Lesley Rogers.
p. cm. — (Maps of the mind)
Includes bibliographical references and index.
ISBN 0–231–12010–9 (cloth : alk. paper)
1. Neuropsychology. 2. Brain—Sex differences. 3. Sex differences (Psychology)
I. Title.   II. Series.

QP360 .R628 2001
612.8′2—dc21   00–060255

Printed in the United States of America

c 10 9 8 7 6 5 4 3 2 1

Dedicated to Gisela Kaplan

Contents

Acknowledgments

I am most grateful to Professor Gisela Kaplan for making detailed and valuable comments on the manuscript, and to Professor Steven Rose for inviting me to write the book and also for commenting on the manuscript. Peter Tallack edited the book and made many suggestions to help improve style and content. The figures were prepared by Craig Lawlor. I thank all of these people.

Sexing
the Brain

New Methods, Old Ideas

This book is about the science of sex differences. It is also about social attitudes and prejudice. The topic of sex differences is part of an interesting debate that has a very long history. Science has contributed to that debate, sometimes with evidence and sometimes in a distorted way reflecting social and political opinion rather than any form of objective truth. Countless studies have been carried out to find out how different women and men really are, and often the differences have been exaggerated while the similarities have been ignored. The differences refer to average differences between women and men, but there is a large amount of variation among women and among men, so there is always overlap between the sexes. Nevertheless, considered in terms of averages, sex differences do exist and we are justified in asking how they have come about and how they develop.

Do women and men learn to be different from each other, or is our biological inheritance mainly responsible for the differences? This question has intrigued us for decades, and it is just as important today as it ever was. The answer, in reality or what we believe it to be, underlies the great social debate about equality of the sexes. If the genes—our biological inheritance passed on from one generation to the next—are the sole or main cause of sex and gender differences, then any social change that might remove the differences between women and men would be merely covering them up. The same inequalities could reappear at any time. But if sex differences are learned and the genes play little to no role beyond causing the physical differences between the sexes and reproduction, then changing the social environment in which children grow up and adults live could remove differences or gender, once and for all.

Whether we consider nature (genes) or nurture (experience and learning) to be more important has great social, political, and economic outcomes. In the recent past, people have tended to take up one or the other extreme position, some people believing that genes have a preeminent role, and others that social or environmental factors are overwhelmingly important. This action and reaction stirred heated exchanges and generated debates that challenged the very basis of society. At the same time, the nature-nurture debate stimulated a great deal of scientific research and demanded greater rigor in interpreting the results obtained, but did not always receive it.

Today, the debate no longer takes quite such flamboyant extremes, but it is essentially still with us and is still of considerable social importance. Although we now realize that all aspects of behavior rely on both experience and genetic contributions interacting in complex ways,[1] many people are still inclined to interpret evidence of biological differences between the sexes as meaning that the differences are hard-wired and fixed by genetic inheritance. Our biology includes our genes and their influences, but it is shaped by experience. There is ample evidence showing that experience can change the biology of the brain (as well as other parts of the body), and learning is an example of this. We should therefore not think of biology and experience as opposites. Genetic influences can be seen as distinct from experience (nature as opposed to nurture), but biology is common to both of these influences on an individual's development.

Nevertheless, these points tend to be overlooked, and differences in biology are often seen as part of the great plan of nature and to be quite immutable. Used in this way, the term *biology* is being equated with *genetic* and so refers to developmentally specified influences on the individual. For example, differences in the way that brains are "wired up," or in the molecular processes that occur within them, are seen to be caused by the genes alone. As we will see, the existence of biological differences between individuals, or groups of individuals (for example, women and men), does not prove that these differences have been specified only by the genes.

Let's be clear about what we mean when we talk of genes. They are lengths of DNA found inside the nucleus of each cell in the body, copies of which are passed on from one generation to the next in the egg and sperm. Each gene has a code for making a particular protein that is made inside the cell; for example, a single gene encodes a protein that influences eye color, and another one does the same for hair color. The code is much more complicated for other physical features and for the structure and functions of parts of the brain. Not surprisingly, then, there are many thousands of genes. These are strung together into chains called chromosomes, which can be seen by looking at cells under the microscope when one cell is dividing into two. It can be seen at this time that each cell in the human body has forty-six chromosomes, which come in pairs matched for shape. The two chromosomes of particular interest to us here are the X and Y chromosomes, so called because they are shaped like these letters of the alphabet. These are the "sex chromosomes" and influence the development of female and male physical characteristics. Women have two X chromosomes, whereas men have one X and one Y chromosome.

Genes located on the X and Y chromosomes are often assumed to influence the development of sex differences in thinking and behavior. They are said to exert their effects by way of the sex hormones, particularly testosterone, estrogen, and progesterone. A chain of influence is considered to occur from genes to hormones to brain structure and function and then to behavior. Many people, including scientists, think of this chain of events as the only cause of biological differences between women and men, and pay little or no attention to influences from outside the individual (such as the influence of culture). But finding a biological sex difference does not tell us anything about what caused it. The assumptions we make about the causes of such differences are, to a large extent, a reflection of social attitudes and have been shaped over a long history.

This book will be critical of simplistic genetic and hormonal interpretations of sex differences. I will argue for a more complete picture taking into account the interactions of genes, hormones, and experience.

In this chapter I will briefly look at the history of thinking about sex differences and then discuss some of the latest technological methods for studying sex differences in the brain. The results obtained using these new technologies will be scrutinized carefully. It is important to do this because any research showing, or claiming to show, sex differences in the brain or behavior affects how we view ourselves and reinforces certain social and economic divisions between women and men.

Before starting, I need to clarify some terms that will be used to refer to the differences between women and men. To begin with, should we use the term "sex differences" or "gender differences"? Some scholars have seen it important to make a distinction between biological sex and cultural gender. As a result, feminist writers have a strong preference for using "gender" instead of "sex" because this emphasizes the importance of cultural influences on the development of femininity and masculinity. By doing so, they focus attention away from the physical aspects of the body and challenge the biological naturalness of differences between women and men. Their aim is to highlight the importance of conditioning and learning in the development of differences between women and men. Researchers in the biomedical field, on the other hand, refer to physical and behavioral differences between women and men as "sex differences." Here the emphasis is on biology and its role in determining the behavioral differences between the sexes.

In the biomedical field, "gender" has a specific meaning stemming from the concept of "gender identity." The gender identity of a person is considered to be the sex that she or he feels herself or himself to be, irrespective of biological sex. An individual may, for example, feel that she is a man even though her physique (and aspects of her biology) is typical of a woman. Or a person who is physically a man may feel that he is really a woman. These people are called transsexuals or, more recently, transgenders. The sex versus gender issue here is completely entangled, as biological sex and gender identity are not consistent with each other. But the gender identity felt by an individual may not be expressed in that individual's general behavior. It is perfectly possible for someone with a woman's biology (sex) to have a male gender identity and still behave in a feminine way (that is, express a female gender role).

Sociologists and feminists usually use "gender" to refer to the expressed behavior (the gender role), rather than to inner feelings of gender identity. At the same time, they recognize the plurality of behavioral expression within the concept of gender. To them, gender is variable and responsive to change, whereas sex is not. Sex is equated with biology and gender with behavior, but this distinction between the two is not absolute. As already mentioned, the idea of biology as immutable is largely incorrect: learning and experience can affect the biological, as well as the behavioral, differences between women and men.

With these problems of expression in mind, I have decided not to avoid referring to "sex differences." This seems to be appropriate because I will be discussing the roles that genes and hormones play, or do not play, in the expression of sex differences in the brain and behavior. There will be numerous places in the text where "gender differences" might be equally, or even more, appropriate, but I will leave that to the reader to decide. It might be even better to refer to a "sex-gender system" because this incorporates the complex interactions between the two but, as this can be somewhat cumbersome, I will stick with "sex differences." I will be exploring biology and its expression in behavior but, while recognizing the importance of terms used to refer to the biology and behavior of women and men, do not want to be constrained by conventions, either new or old. At the same time, the chapters to follow will discuss the social and political implications of research on sex differences and the way in which the results are interpreted.

Old Ideas

The predominant popularity of genetic and hormonal explanations of human behavior reflects the fact that interpretation of results is not "value free." How we interpret results is surreptitiously influenced by a host of assumptions and views that have a very long history. That history has been biased against women. There is a great deal of evidence in the scientific and popular literature, extending well into this century, that the mental capabilities of women have been considered inferior to those of men. These attitudes have influenced what scientific investigations have taken place and what interpretation of the results has been seen as acceptable. This is why we need to know something of the history of the ideas that have surrounded research on sex differences.

As a scientist, I must keep pace with the new knowledge of my discipline, but I should also know how thinking in the past has molded current ideas and might influence how we interpret any new findings. Philosophy aside, debates about human nature have generally been about the role of biology in determining human behavior, so those aspects of biology that might contribute to understanding human behavior have held a special place in the discipline of biology as a whole. No other area of biology is more influenced by social attitudes than the study of differences between human groups.

Last century, it was thought that women lacked the region of the brain in which "the intellect" was said to be located. It was popular to draw attention to the fact that, on average, women's brains weigh 10 percent less than men's. As late as the end of the nineteenth century, this fact was used as a reason to oppose improvements in the education of women. The sex difference in brain size was used to justify sexual inequality. The science of craniology—measuring the size of the cranium—flourished at that time. Among the craniologists was G. LeBon who, in 1879, wrote:

> In the most intelligent races, as among the Parisians, there is a large number of women whose brains are closer in size to those of gorillas than to the most developed male brains. This inferiority is so obvious that no one can contest it for a moment; only its degree is worth discussion. All psychologists, as well as poets and novelists, who have studied the intelligence of women recognize today that they represent the most inferior forms of human evolution and that they are closer to children and savages than to an adult, civilized man.[2]

This and similar negative views about the mental capabilities of women were linked to attitudes toward non-European ethnic groups, usually of a different skin color—the so-called "savages" who were seen to require civilizing.[3] The claimed inferiority of women was linked to the claimed inferiority of those ethnic groups that Europeans were dominating through colonization. In 1871, Charles Darwin wrote that at least some of the mental traits in which women excel are traits characteristic of the lower races.[4]

Measurements of brain size relative to body size made these notions obsolete by around the turn of the century. As Franklin Mall demonstrated in 1909, there is no difference between the sexes when brain weight is adjusted for body size.[5] Mall also demonstrated that there is no sex difference in the number of convolutions of the cerebral cortex (the folds visible on the surface of the cortex), another feature that had been associated with intelligence (see fig. 1.1). Others had concentrated on the relative sizes of parts of the brain in women and men. At first, in the middle of the nineteenth century, it had been thought that the frontal lobes were on average larger in men than in women. Later, toward the end of the nineteenth century, it was asserted that the frontal lobes were on average smaller in men than in women but that the parietal lobes (at the back of the brain) were on average larger in men than in women. At this point, many scientists immediately switched their argument to say that larger frontal lobes do not indicate superior intelligence, but that the parietal lobes were more important.[6]

Despite the lack of anatomical evidence to support the notion that the female brain was inferior to the male brain, the underlying theme behind the thinking of the late nineteenth century persisted in different forms. In the first quarter of the twentieth century, Havelock Ellis formulated the theory of "greater male variability." He believed that, whatever characteristic one measured, there would be more variability in a group of males than in a group of females. Much later, in the 1970s, this idea was applied to the interpretation of tests of intelligence (IQ tests) in men and women. Greater variability for the mean (or average) in men was said to explain why there is a higher proportion of men amongst the great artists, scientists, musicians, and so on.[7] More men than women were said to have very high IQ scores. Indeed,

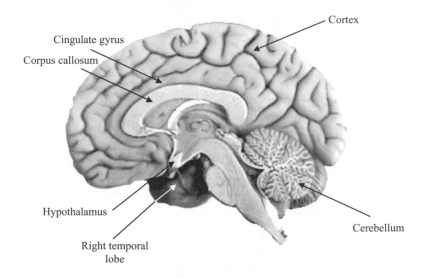

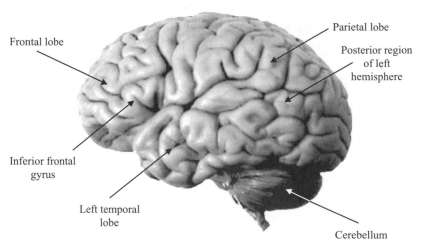

FIGURE 1.1 Two views of the human brain. *Top:* The right half of the brain after it has been sectioned through the middle. The view is from the midline, at the cut side. *Bottom:* The left side of the brain, viewed from the outside. In both cases, the person would be facing to the left (that is, the front of the brain is on the left and the back of the brain is on the right).

more men were said to have very low IQ scores too, although the low range received little attention. The scores of women were closer to the average than those of men. This is another way of expressing female inferiority, now in terms of their sameness or being more average than men.

But the reason for the sex difference in distribution of IQ scores is, to a considerable extent, a matter of how the tests are constructed, rather than a reflection of any basic difference between women and men. Women and men do perform differently on various questions (spatial versus verbal ones) in an IQ test, but the overall IQ result depends on the balance of questions in the particular test, rather than being a manifestation of something in their biological natures. There have been IQ tests on which women originally scored higher than men, but these have been adjusted to eliminate the female superiority. There are many different IQ tests, and the differences in the distributions of overall IQ scores for women and men vary from one test to another. There is also evidence that IQ tests are not as independent of past experience and social position as is often claimed. This might also explain at least some of the differences in the scores of women and men. To dwell on the sex difference in the distribution of IQ scores, and in particular to draw attention to the high end of the scale to explain why more great scientists, artists, musicians, and so on are men, is not value free but is an aspect of negative attitudes toward women.

In a less obvious way, negative attitudes about women continue to underlie a broad spectrum of research into sex differences today. Claims that the differences are caused by genes or hormones confirm beliefs in the inferiority of women and resist change to equality across the gender divide. Genetic theories for the cause of sex differences in behavior attracted new interest during the 1970s when women began to demand equality in the home and at work. The genetic determinist position was part of the backlash to the growing feminist movement of the time. Many writings by feminists then argued that genetic inheritance has little to do with generating the differences between women and men, and that these are instead determined by learning in a culture that treats women and men differently.

In reaction to the feminist position, some endocrinologists and

psychologists said that the behavior of women is influenced by the ebb
and flow of their hormones as they change during the menstrual cycle,
or during pregnancy.[8] Irrespective of the particular skills of a woman,
during a certain stage of the menstrual cycle her ability to exercise those
skills would be diminished, according to this viewpoint. It was therefore argued that even a woman with exceptional mathematical skills
would be a liability for employment in a responsible position because
there would be monthly periods in which she would be unable to exercise those skills.[9] A similar argument was used to prevent women from
becoming airline pilots. At the time, these views received wide coverage and support in the popular media.

Belief in genetically or hormonally caused sex differences in ways
of thinking or reasoning is still used to divide the sexes socially and in
employment, as shown by the following quote from 1992 by Doreen
Kimura of the University of Western Ontario:

> The finding of consistent and, in some cases, quite substantial sex
> differences suggests that men and women may have different occu-
> pational interests and capabilities, independent of societal influ-
> ences. I would not expect, for example, that men and women would
> necessarily be equally represented in activities or professions that
> emphasize spatial or math skills, such as engineering or physics. But
> I might expect more women in medical diagnostic fields where per-
> ceptual skills are important. So that even though any one individual
> might have the capacity to be in a "nontypical" field, the sex pro-
> portions as a whole may vary.[10]

As will be discussed in detail later, Kimura has published results
showing that the areas of the brain used for speech may be located at
different sites in males and females. She has not conducted any studies
to see what might cause these differences. Nevertheless, she assumes
that they are not caused by social influences but by the action of biological factors, such as genes and hormones. She goes on to say that the
unequal representation of women and men in professions requiring
spatial skills, in her view, is due to the "fact" that women supposedly
have inferior spatial abilities to men. This, she believes, is a reflection

of the biological differences between women and men rather than an aspect of social inequality and a result of the social perceptions that perpetuate unequal representation of women and men in the professions.

This is a rather small selection of examples illustrating the background to research on sex differences, but there are many more along the same lines. Beginning in the 1970s, some feminists, some of whom are scientists, have tended to deny that biology (genes or hormones) is important in sex differences. Some scientists have responded with equally strong assertions that the influence of biology is paramount. From this debate has emerged an active interest in research on sex differences in both biology and behavior in animals and humans. This interest has persisted to the present day.

New Methods

Current debates about biological contributions to gender differences take shape around the new technologies that allow us to probe the brain and reveal how it thinks and functions. We can now obtain computer images that show the activity of nerve cells in the living brain, and this allows us to see what parts of the brain are active when a person performs a particular task or thinks about certain things. Images on a computer monitor show the most active regions of the brain in reds and yellows and the least active regions in blues and greens. Fine measurements can be taken of the sizes of structures deep within the brain and of the levels of the sex hormones circulating in the bloodstream, as well as in the brain itself. Other methods make it possible to probe the human genome to identify genes that affect the development and function of the brain.

These new techniques are aimed at understanding brain function in a broad range of ways, including functional abnormalities. For example, imaging of the activity of the brain of someone who has had a stroke can provide valuable information about the damage that has occurred, as dark areas in the image show where the blood supply has been cut off or where the nerve cells are inactive. Using the same procedure on someone suffering from epilepsy can reveal where the seizure starts and follow the spread of the changes in activity as it progresses; in this case, we are seeking red spots of high activity. There are many other direct medical applications of the new procedure. They are used in the hope of diagnosing medical conditions and finding new therapies.

It was no surprise when these technologies for examining the brain were applied to study sex differences in brain activity, structure, and chemistry. Sex (or gender) differences always fascinate humans and, as we have seen, there has been a growing interest in the area. But unlike the results of research on abnormalities in brain function, the results of research into sex differences will not find a cure, although it is useful to know of baseline differences in "normal" female and male brains for certain types of therapy. Instead the search is largely for new knowledge. There is nothing wrong with this in itself, but perceived biological difference is usually used to divide one group of people from another. All too often one group describes itself as "biologically" superior to another to justify holding power or most of the resources in a society. Biological difference is rarely seen as being value free. The now substantial amount of research on comparing the fine structural aspects of the brains of heterosexuals and homosexuals, or heterosexuals and transsexuals, is not simply a matter of scientific interest. It has social implications. Heterosexuals are considered by a large part of society to be superior to homosexuals and transsexuals, and biological differences are used to justify this position.

Women versus men and homosexual versus heterosexual are just two of the many ways in which we can divide people according to social criteria. It would be possible to look for differences in the brain that might correlate with any one of a number of divisions made according to social criteria. Not so long ago, it was considered to be of scientific interest to make similar comparisons of people divided along the lines of presumed race or class. For obvious reasons, such research is no longer entirely acceptable, and hopefully it never will be again. But the search for sex differences in the brain has never gone out of fashion, although interest has tended to wax and wane. The same is true of the search for the supposed biological bases of homosexuality and transsexuality. These trends are reflections of social attitudes. When society wants to maintain inequality, biological explanations can be used to justify it.

Although scientific methods are used in researching sex (or gender) differences, the decision to carry out this research in the first instance has little to do with science. From its inception, this research is never value free. We should remember this when the results of such research are interpreted and fed back to society to provide a framework for future social decisions. The findings do not just fulfill scientific curiosity but also serve social and political purposes far beyond the boundaries of science. This point is starkly apparent in the scientific research comparing the brain structure and function of homosexuals and heterosexuals, or transsexuals and heterosexuals. The results of these studies have been used as a basis for legal sanctions against the practice of homosexuality and transsexuality, and for the medical treatment of homosexuals and transsexuals. I will discuss this further in chapter 3.

The interest in sex differences and related topics might stem from our ubiquitous interest in sex itself, but sex as an activity has little to do with the broad aspects of behavior of women and men, and their daily lives at work and home. Our interest in sexuality, it seems, is not the driving force behind our interest in sex differences in behavior. Instead, that interest is stimulated by competition for power, both social and economic. Inferiority versus superiority in this context is, by and large, demarcated by sex. Women are not just different from men; worldwide, women still hold less economic power than men and, in general, have a lower social status. This power imbalance often gives impetus to research on sex differences, and sexual orientation, and it drives research to the extent that vast amounts of funding are made available for investigating differences in behavior, mental ability, brain activity patterns, and brain structure.

Perhaps the prime example of this new and costly research on sex differences is the mapping of brain activity using positron emission tomography (PET) and functional magnetic resonance imaging (fMRI).[11] These techniques are revolutionary because they allow neuroscientists to scan the living brain and see patterns of activity change as the brain performs different mental tasks.

PET makes use of the fact that the areas of the brain in which the nerve cells are more active have increased blood supply. Radioactive oxygen or glucose is injected into the bloodstream and carried in the blood to the brain. It is common to use a form of glucose called deoxyglucose because, unlike glucose, it is not converted to any other substance after being taken up into cells. In this way the radioactive molecules accumulate in cells in the most active regions of the brain. They emit positrons, which are small particles with a positive charge. The positrons collide with electrons, which are negatively charged. These collisions produce gamma rays, which penetrate through the brain and skull and are detected by scanning equipment placed around the subject's head. This information is fed into a computer program that generates an image of the brain on a monitor screen. The image shows which areas of the brain are most active, and which areas are least active.

fMRI reveals the activity of the nerve cells by measuring the amount of deoxyhemoglobin, which is a form of the protein hemoglobin, in the blood flowing to the different regions of the brain. In this case there is no need to inject anything into the bloodstream. Hemoglobin in a form called oxyhemoglobin carries oxygen in the blood. When the oxygen is supplied to active cells, deoxyhemoglobin forms. Because the magnetic properties of the hemoglobin are affected by whether it is oxygenated or deoxygenated, it can be used to discover those areas of the brain that are more active. When an area of the brain becomes active, its blood supply increases and the level of oxyhemoglobin increases. This changes the magnetic field locally. In this way, a pattern of activity can be generated on the computer monitor.

By scanning the brain, computerized images are generated that show the patterns of activity at different sites in the brain. Some even newer methods of scanning detect the levels of neurotransmitters, which are the substances used by nerve cells to communicate with each other. A scan gives a cross-section of the levels of activity in, for example, the different hemispheres of the brain. It allows us to see which areas of the brain are active when a subject is speaking, reading, listening to spoken words, solving a mathematical problem, or performing any other mental process, provided that it can be done while the subject is lying still with his or her head inside a tubular apparatus. The subject can perform limited motor tasks with the hands, such as tapping rhythms, but not anything that would lead to movement of the head.

The need to have the subject's head in the scanning apparatus could be a major drawback, not only because it causes claustrophobia, but also because it is possible that the brain functions very differently when the subject is able to move around and does not have to concentrate on keeping the head still. Unfortunately, we cannot find this out with the available technology, so this potential flaw is ignored. Another problem is that the magnetic sources in fMRI produce a continuous loud noise that is heard by the subject. This could interfere with the processing of auditory stimuli and even those stimuli detected by the other senses. These problems may well affect the results of such research, but they are rarely discussed in papers reporting results obtained by brain scanning.

Sex Differences in the Brain

Richard Haier of the University of California, Irvine, and his colleague Camilla Benbow have used brain-scanning technology to obtain results showing sex differences in brain function.[12] Young women and men were PET-scanned while they were solving mathematical problems. Before the experiment, each subject was categorized as either a high or an average performer on the basis of tests in mathematics. Each subject was injected with radioactive deoxyglucose and, for the next half hour, during which the deoxyglucose was taken up by active cells in the brain, the subject was asked to perform mathematical tasks. Immediately after this, the PET scan of the subject's brain was carried out. A sex difference was observed in the high-scoring group: the men had more activity than the women in the temporal lobes (at the sides of the hemispheres of the brain; see fig. 1.1). Other areas of the brain were also active, but these were not found to differ in the women and the men. Neither average-scoring men nor average-scoring women had this higher level of activity in the temporal lobes.

Women and men with similar high-performance scores were therefore using different regions of their brains to perform the mathematical tasks. This might suggest they solved the problems using different strategies and therefore different regions of the brain hemispheres. But, although the high-scoring men may have been using their temporal lobes for mathematical reasoning, they could also have been using them for other things at the same time. In other words, the evidence is not conclusive that the subjects used the temporal lobes for solving the mathematical problems. It is possible to think about more than one thing at a time, and they may have used their temporal lobes to process information about events going on in the room at the same time as they were solving the mathematical problems. The women and men might have been thinking about, or giving attention to, different things during the tasks. They might also have different degrees of ambivalence to or interest in the experiment, or they might have been more or less stressed by the procedure. Such confounding inputs to the results could have been established just after the subjects had been tested by asking them some simple questions or by measuring heart rate or other responses to stress during the test itself, but this was not done.

In the men, there was a mild but statistically significant association between the amount of deoxyglucose taken up by the cells in the temporal lobes and the number of mathematical problems that the subjects solved correctly during the experiment. At first glance, this result might seem to confirm that the temporal lobes are the place in the brain where mathematical problems are solved, as concluded by the researchers. But this is not the only explanation. Perhaps the activity in the temporal lobes reflects more motivation to perform well on the task. If so, a correlation between temporal-lobe activity and the number of problems solved correctly would not mean that the problem-solving is located in the temporal lobes. I have stressed this point to illustrate that a correlation between two factors does not mean that they are causally related to each other. They could be unrelated but vary together because they are both controlled separately by a common third factor.

Another study using PET-scanning reported sex differences in the pattern of activity in brains "at rest." Ruben Gur and colleagues at the University of Pennsylvania asked the subjects to rest while the PET scan was taken.[13] Glucose metabolism, indicating nerve-cell activity, was higher in the temporal limbic system (a number of regions in the temporal lobes) of males than it was in the same region in females. This system is one of the parts of the brain that plays a role in the emotions. But glucose metabolism in the cingulate gyrus was higher in the females than in the males. This is another region of the limbic system, linked to emotional responses but perhaps at a higher level of processing. The cingulate gyrus is also associated with perception of pain, and is involved in the relay of higher mental processes of the cortex to lower regions in the hypothalamus. Gur and colleagues interpreted their results as indicating that sex differences in cognitive and emotional processing have a biological substrate. But this conclusion reflects beliefs rather than findings. For any number of reasons, women and men may think about different things when they are at rest in the particular context of the experiment. This might depend not so much on having basically different biology, but on differences in past experience.

This is not to say that sex differences in brain function do not exist. There is some evidence, for instance, that language may be processed in different parts of the brain in women and men. Bennett Shaywitz of Yale University, together with many colleagues, took fMRI images of the brains of male and female subjects while they performed a number of tasks using words.[14] One of these tasks revealed a sex difference in the images. When the male subjects were asked whether two strings of nonsense words rhymed or not, nerve-cell activity was higher in a region in the left hemisphere known as the left inferior frontal gyrus (fig. 1.1). When the female subjects were rhyming words, the activity was higher in both the left and the right inferior frontal gyrus than in other parts of the brain.

This apparent sex difference in the fMRI images could be criticized on the same grounds as mentioned previously for the PET scans, but the bias to the left hemisphere in men and not in women had been indicated by some previous research. This had shown that women use regions of both the left and the right hemispheres when processing language, whereas men use the left hemisphere more than the right. This had been discovered by flashing words on screens located to the left and right of the subject. The subject had to look straight ahead so the words were visible at the edge of the left or right field of vision, which ensures that the left hemisphere processes the words seen in the right field of vision and the right hemisphere processes the words seen in the left field of vision. The subject was asked to read the words. Performance in both men and women was better when the right visual field (and the left hemisphere) was used than when the left visual field (and the right hemisphere) was used. But women performed better than men when the words were presented in the left visual field.[15] This result indicates that women may process some aspects of language in the right hemisphere as well as the left, and the fMRI images obtained by Shaywitz and colleagues are consistent with this.

So there seems to be a sex difference in the use of the left and right sides of the brain for language processing. There may also be a sex difference in the use of the front and back regions of the brain hemispheres. In the early 1980s, Kimura worked with subjects who had aphasia—that is, they were unable to produce speech—following brain damage caused either by a stroke or a brain tumor.[16] She located the damage using electroencephalographic recordings.[17] This is not the best method for finding the damage (today fMRI or PET-scanning is more accurate), but in broad terms she found a sex difference. She reported that women suffering from aphasia were more likely than the men to have damage in the frontal part of the brain. In aphasic men, the damage was more likely to be in the back part of the hemispheres. But these results could have been compromised by the fact that more of the women in the study had tumors than did the men, a larger proportion of whom had damage caused by a stroke.

In a subsequent study in 1989, Andrew Kertesz of the University of Western Ontario and Thomas Benke of the University Clinic, Innsbruck, Austria, examined only stroke patients with aphasia and found no sex difference in whether the site of damage was in the front or back regions of the hemispheres.[18] They used computer imaging to locate the damage, so their study was considered to be more accurate than that of Kimura. This result confuses the issue of whether there is a back-front difference in where women and men process language in the brain. Kertesz and Benke also found no evidence of a sex difference in processing of language in the left or the right hemisphere. So we can say that women and men certainly use the same primary regions of the brain for producing speech and processing language (these being in the left hemisphere in most people). We can also say that there are indications, but no more than that, of a sex difference in the use of some additional regions of the hemispheres for language processing.

These are just some of the many studies that have reported sex differences in brain function. Even more studies have reported sex differences in ways of thinking. They report that women, on average, have verbal and linguistic skills that are superior to those of men, and that men, on average, have better spatial ability than women.[19]

Many of these studies have used large numbers of subjects, for only then do some of the sex differences emerge. This is because the differences in the average scores for women and men are actually very small, and there is a large amount of overlap between the scores obtained for women and men. It takes a large number of subjects to show up small differences in the averages and, in such cases, it could be said that the overlap between the two groups is as important as the small difference in the averages. This is the way in which similarities of the mental abilities of the sexes are overlooked. People tend to focus on the differences in averages. This has happened to such an extent that some people have seen it timely to draw attention to similarities rather than differences.[20] This is certainly a valid stance to adopt, and I have much sympathy with it. But, even though the sex differences in brain-cell activity and mental style might pertain only to averages obtained by testing large numbers of subjects, it is not unreasonable to question how these differences in averages might have come about. Given the increasing amount of literature reporting sex differences, it is important to pursue the answer to this question, rather than avoiding it.

How Sex Differences Develop

In the following chapters, I will address, as a biologist and a feminist, why these sex differences might exist, and therefore how malleable they might be from one generation to the next. I am not inclined to take the fallacious line of least resistance and say that evidence for differences between female and male brains implies that they are caused by the genes. Also, I am not prepared to say that the sex differences in humans are a result of evolution, as some have done. They have argued that better performance by men on some spatial tasks stems from an evolutionary past in which males might have needed to use spatial abilities to defend territory and hunt for food. This fanciful thinking is not, I am afraid, absent from science today. I will discuss this further in the next chapter.

So why are some scientists, and nonscientists too, so ready to believe that any sex difference in brain structure or function must necessarily imply a genetic cause, with little or no influence of learning? There is no particular reason to break with the scientific maxim of keeping an open mind until the point is proven by controlled experimentation. Yet we seem unprepared to do that when it comes to sex differences. Something seems to drive us to assume that these have genetic or hormonal causes. Recent advances in molecular genetics and the vast amount of money poured into this research make it all the more tempting to attribute a greater slice of influence to genes than to other factors. We might just as inaccurately assume that sex differences are caused entirely by environmental influences, but this has been a far less popular view.[21] There is a long-standing history of belief in genetically caused, fixed differences between women and men, and this belief has been applied, both openly and surreptitiously, to justify the inferior status of women in society.

We will begin to understand the causes of sex differences in brain structure and function only by carrying out well-controlled experiments that take into account all of the factors that could affect the structure and function of the brain. This means that researchers will have to consider potential influences of genes, hormones, and experience. Although this will make any investigation of sex differences much more extensive, and therefore more difficult to carry out, than most of the current investigations, the results of such studies are more likely to be reliable. Researchers who tackle the question of sex differences will need to be conversant with more than one field, and they will be asked to prove, rather than assume, what might be the cause of any differences they discover. Without doubt, the results of such investigations will reveal interactions among genes, hormones, and experience. The development of an individual requires change, and change is the product of interacting influences, some of which come from inside and some from outside the individual.

Before going any further, I need to discuss the process of development, because it is essential to understanding the causes of sex differences. It is common to think of an individual organism, be it a human, an animal, or even a plant, as beginning life with some sort of basic program encoded in its genes, and that learning (or experience) modifies this program as the organism matures. The program is seen as being within the organism and the experience that modifies it comes from outside the organism. In this way, we conceive of some sort of action and reaction between factors inside and outside the organism. But this is not really an appropriate description of the developmental process. The inside and outside mutually interact.

The genetic program seems to set certain conditions on what will be learned, or, to put it in a different way, on what kind of experiences will leave their mark on the individual. This way of conceiving the process of development is very important when it comes to sex differences. As we will see, recent explanations of the cause of sex differences do consider learning to be important, but only within the constraints of the genetic program. Put simply, according to this view, girls might learn to be girls, and boys might learn to be boys, but the genetic program will ensure that girls learn the kind of things that make them girls, and boys learn the kinds of things that make them boys. But how much evidence is there for this view?

We know the genetic program does determine what things can be learned, but only within a broad context. For example, a species with a superior sense of smell will learn more about the odors in the outside world than a species with a poor sense of smell. There are numerous examples of the genetic program guiding very general aspects of learning, but whether it makes sure that girls learn to be proficient in language and other things associated with femininity, but not in mathematics, is unlikely. These are more detailed and complex differences, and there is no evidence that the genetic program for development channels behavior to this extent. A program that might do so would require the expression of an enormous number of genes in very complex ways, but this does not seem to happen.

There are sensitive periods during development when certain things are learned easily and rapidly. The timing of these special stages in development can be programmed in the genes, but exactly what is learned during each sensitive period is not part of the program.

Although girls may develop at a different rate than boys, and therefore pass through sensitive periods at different times than boys, this need not mean that they learn "girlish" things at those times. Although it is important to recognize the interactions between the genetic program and learning during development, it is also important to note that the genetic program sets only certain broad guidelines on when certain experiences will have an effect and on what will be learned. The entire process of an individual's development is not merely a battle between nature (genes on the inside) and nurture (experience or learning on the outside), but is a dynamic interweaving of processes within a system which is inseparably the organism and its environment.[22] The behavior and social position of an individual is not the result of a preexisting plan shaped by experience. Rather, it is an outcome of interactive processes occurring at all stages of development and always subject to further change. Genes are neither central nor primary to the developmental process, and the same is true of the effects of learning.

What Causes Sex Differences?

Although it might be interesting to use psychological tests to find out whether sex differences exist, the mere fact that they do exist tells nothing of their origins. Despite the new ways of measuring various aspects of brain structure and function, we seem to have made little progress toward a genuine understanding of what really takes place as a person develops. We know little about what causes an individual to develop behavior patterns largely associated with one sex and not the other. In other words, although we can measure differences between the brains of women and men, and the list of differences might even grow as new measuring techniques become available, this knowledge has shed little light on the causes of these differences.

The discovery that females process language in a different part of the brain than males tells us nothing about what caused this difference to develop. It is much easier to identify differences between the sexes than to find out what caused them. Part of the problem lies in the fact that, in all human societies, women and men have differences in both their biology and their social environments. It is virtually impossible to separate the contribution that each of these two variables might make to the measured differences between women and men.

Girls and boys are raised differently almost from the moment they are born. Social differences are often considered to result "naturally" from biological differences, but this need not be the case. In fact, it can be the other way around: many biological differences may result from the influence of being raised in different social environments. The environment first becomes different when baby girls are dressed in pink and baby boys in blue. This may continue by mothers speaking more to girls than to boys, and encouraging (or discouraging) girls and boys to play with different toys and to show different amounts of aggression. All too often, biologists studying sex differences ignore such effects of social experience on biology, or at least underplay them. They make no attempt to control for the influences of social experience on their subjects, and they often say nothing about it when they publish their results. This omission may arise because most scientists who study sex differences have little to no training in such fields as social psychology that would increase their understanding of these issues. Also, the potential effects of social experience are so complex and difficult that some may consider it easier to ignore them.

All of the information available about sex differences in brain function is obtained by comparing female and male subjects. This approach tells us what sex differences exist, but it cannot tell us what caused them. To find out what caused a particular sex difference, we would have to do an experiment in which we changed a factor that we considered to be important, and then see whether it affected the sex difference. For example, to find out whether sex hormones affect the development of the region of the brain used when rhyming words, we would have to alter the hormone levels of the child and see what long-term effect this has on the pattern of brain activity when the subject performs a rhyming task. This kind of experiment cannot be done because it would be entirely unethical.

As an alternative, we might make use of "natural" experiments by testing subjects known to have been exposed to unusual levels of a sex hormone during early life. Perhaps we find subjects who secrete higher than average levels of sex hormones. Alternatively, they may have been exposed to an abnormally high level of a hormone while they were developing inside the womb (*in utero*) because their mothers secreted abnormally high levels of it, and because maternal hormones can reach the fetus through the placenta. Or maybe their mothers were treated with hormones for some medical reason during pregnancy, such as to prevent miscarriage. We could then look at the pattern of brain activity in females who had been exposed to unusually high levels of male sex hormone before or just after birth. Alternatively, we could look at individuals raised in ways that are not typical for their biological sex to see whether social and psychological factors might have contributed to the difference. Depending on the results obtained, this might tell us something about the factors that were instrumental in causing the sex differences.

But it is difficult to obtain suitable subjects for such natural experiments. This is the main reason why experiments aimed at discovering what causes sex differences are carried out far less often than those merely documenting the differences between women and men, or girls and boys, who fall within the average range (that is, with no known hormonal abnormality during development). Such studies provide no scientific evidence for what causes differences. Surprisingly, however, researchers are usually inclined to give causal explanations for their results. But we must keep in mind that in doing so they are making assumptions or stating beliefs, rather than drawing on evidence. The claims about what causes the differences extend beyond the factual information obtained.

So what factors could lead to the development of sex differences in brain function and cognition?

The Biological Candidates

The biological factors that might influence the sex differences in brain function are genes and hormones. As we saw in chapter 1, genes are the hereditary material found in chromosomes inside the nucleus of cells, and there are two chromosomes that discriminate between the sexes, called the X and Y chromosomes. Females are characterized by having two X chromosomes, whereas males have an X and a Y chromosome. We say that females are XX and males are XY.[1] Both sexes have another forty-four chromosomes (in twenty-two pairs) that do not differ between the sexes.

Genes located on the X and Y chromosomes influence the development of the gonads (the sex glands) so that they develop into either ovaries or testes. Once they have become either ovaries or testes, the gonads secrete particular sex hormones that help to cause the development of the physical (but not necessarily the behavioral) characteristics that distinguish the sexes.[2]

Usually males have one X chromosome and one Y chromosome in their cells, but sometimes they have two or even more X chromosomes as well as the Y chromosome (that is, they are XXY or XXXY). An individual with this genetic makeup is still male in appearance. There may also be more than one Y chromosome (XYY, for example), in which case the individual will also be a male, because, as long as there is at least one Y chromosome present, the gonads will develop into testes. This fact shows us that the Y chromosome plays the main part in establishing the development of male gonads. When no Y chromosome is present, ovaries develop instead. As recently as 1990 it was discovered that there is a particular gene, called the *SRY* gene, which is located on the Y chromosome, that causes the gonads to develop into testes.[3] The *SRY* gene turns on a cascade of other genes, and the expression of all of these genes results in the formation of the testes, although the details of how this happens are not yet known. If the *SRY* gene is inserted into the genetic material of a mouse embryo with female (XX) genes, the mouse will develop testes.

The type and amounts of sex hormones produced by the gonads and secreted into the bloodstream depend on whether the gonads develop into ovaries or testes. In turn, these hormones modify the development of the genitalia. So, through the sex hormones, genes determine whether the structure of the genital organs is female or male.

At puberty, the levels of the sex hormones secreted increase greatly. They also fluctuate over the menstrual cycle and change with aging and in response to inputs from the environment, including stress. We usually associate the hormone testosterone with males, in whom it is produced by the testes, and estrogen and progesterone with females, in whom it is produced by the ovaries, but none of these hormones is limited to only one sex (as we will see later). From the bloodstream, these hormones enter the organs and tissues of the body, including the brain. The hormones build up in some tissues because their cells contain special molecules, called receptors,[4] that match the hormone and lock onto it, keeping the hormone in the tissue. Tissues that have receptors for the sex hormones are targeted by the hormones. The more receptors they have, the more hormone they can take from the bloodstream. This is the case for the genitalia and other parts of the body that make the differences between female and male physique, which are referred to as secondary sexual characteristics. Certain regions of the brain also have receptors for the sex steroid hormones and so are targets for the hormones.

There are many receptors for the sex hormones in a part of the brain called the hypothalamus, which is located beneath the hemispheres of the cerebral cortex and above the pituitary gland (see figures 1.1 and 3.1). The hypothalamus is involved in controlling the levels of the sex hormones secreted into the blood by the gonads, and it does so by influencing the release of other hormones from the pituitary gland.[5] In rats and several other nonhuman species, part of the hypothalamus is known to have a role in controlling sexual behavior. The presence of these receptors and the effects of injecting hormones directly into the brain show that the sex hormones can indeed affect the brain. But do the hormones "sex" the brain? Do the hormones released by the developing gonads act on the brain to cause it to develop ways of thinking and behaving that are typical of a girl or a boy, or of a woman or a man?

The pattern of development from genes to hormones to gonads leads to the sexual differentiation of the genitals. The presence of XX or XY genes establishes the release of particular sex hormones which, in turn, cause the genitals to develop as male or female. Many neuroscientists consider that the same sequential process happens during the development of the brain and cognition (that is, from genes to hormones to brain function).[6] In this view, the sex hormones released by the gonads act on the developing brain to cause it to become either a "female" brain or a "male" brain. Various influences from the environment might modify this differentiation process but not its basic separation of female from male. So, according to this idea, the primary differences in the way women and men think and behave result from XX or XY genes exerting their effects on the gonads. This leads to either the female or the male mix of hormones, which in turn acts on the developing brain to direct its development into a female or male type. Although this simple model for brain development is widely accepted, it has not been shown that the same processes at work in the genitalia actually take place in the brain.

Indeed, it is most unlikely that the brain obeys the same rules of development as the gonads,[7] as the gonads are fairly simple organs, whereas the brain is the most complex organ in the body. There are many more different types of cells in the brain than in the gonads, and the nerve cells in the brain are connected to one another in many complex ways. We know quite a lot about the processes that take place during the development of the gonads, but we know relatively little about the processes that take place in the developing brain.

Although we can study the brain and find out some of the ways in which nerve cells function, and we can study the behavior of a person, it is very difficult to link the two together. As we have seen, we might be able to find out that some aspects of the brain's biological function tend to differ in women and men when they perform a particular task or type of behavior, but that is only a correlation between the biological sex and cognitive function. To know that there is a sex difference in the biochemistry of the brain or the activity of nerve cells, and also that there are sex differences in behavior, does not tell us whether or not these aspects of brain and behavior have a causal link. Even if they are causally related, we are unable to say whether the biological factors cause the behavior or whether the behavior causes the biological factors. So we cannot say whether sex differences in the brain cause the sex differences in behavior (and mental processes), or whether behaving as a male or female causes the sex differences observed in the brain.

The brain responds to inputs from the environment, often by changing its biochemistry, and even its structure. When learning occurs, the connections between nerve cells are changed. Experience, particularly during early life when the brain is developing, can bring about changes in the connections between nerve cells, and even alter the structure of certain regions of the brain quite radically, depending on the nature of the experience. We know this mainly from research on animals. For example, if kittens are raised without visual stimulation, the part of the brain that normally processes visual information is invaded by nerve cells that process auditory information. A whole area of the brain is restructured to deal with hearing instead of sight. Social experience also alters the size of areas of the brain and the connections between nerve cells. As an example of this, a rat living in a complex social environment has a thicker cortex (part of the brain hemispheres) than one raised alone. These examples, and there are many others, indicate strongly that the different social environments in which girls and boys are raised may alter their brain structure and function. This is one of the many ways in which brain development differs radically from the development of the gonads.

Although the gonads do not develop into entirely separate "female" or "male" types (there is some overlap between these two categories), separate male and female types are even less characteristic of the brain. The brain does not choose neatly to be either a female or a male type. In any aspect of brain function that we can measure, there is considerable overlap between females and males. As we have seen, not all men have better spatial abilities than all women, and not all women have better language abilities than all men. In fact, the sharp division between women and men that has been constructed by society is much more polarized than any measured differences between the sexes. Society puts women and men, and girls and boys, into separate categories and constructs absolutes. A person is said to be either one or the other. The similarities are ignored and the differences exaggerated.

Biological Overlap Between the Sexes

When a baby is born, whether it is assigned as "male" or "female" usually depends on whether it has a penis or not. Because the clitoris can sometimes be as large as a penis, incorrect assignments have sometimes been made. Sometimes an XX child may be assigned as male and an XY child assigned as female. Cases of the physical sex and genetic sex being mismatched are not all that rare, and are very disturbing for the family, so some hospitals withhold announcing the sex of newborns until they have been tested genetically. In such cases, the child's sex is assigned according to whether it has XX or XY genes, rather than on the basis of the physical characteristics of the genitalia. Irrespective of variations in the physical appearance of the genitalia, the assignment is a choice simply of either female or male. If the genitalia remain divergent from the genetic sex as the child develops, they are surgically corrected to bring these in line with the genetic sex, making physical appearance conform to the category of either female or male. The biological variation is therefore constrained by our constructed views of what is female and what is male. In the Western world, surgery is now often performed to make the physique of individuals conform to a perceived model of what is male and what is female, as well as what is beautiful and what is not. In most cases, it is women for whom the pressure to conform to the ideal image is stronger than it is for men. This means that the female category is more narrowly defined than the male category.

Just as we assign individuals to the categories of female and male, so we refer to the male and female sex hormones. Testosterone is referred to as a male sex hormone, and estrogen and progesterone are referred to as female sex hormones. As a result, we tend to think that only males have testosterone and only females have estrogen and progesterone. This is not the case. Both females and males secrete all three sex hormones, although they do so from different organs. For example, whereas males secrete testosterone from the testes, females secrete it from the adrenal glands. On average, males have more testosterone than females, but there is a considerable overlap, with many women having higher levels of testosterone in their bloodstream than many males. Similarly, although estrogen and progesterone are referred to as female sex hormones, there are times when men have higher levels of these hormones than women. In women, the levels vary according to the menstrual cycle and, during one phase of the cycle, the levels of both these hormones in the bloodstream of a woman are lower than the levels of the same hormones in men. Also, after the age of about fifty, men have on average higher levels of estrogen and progesterone than do most women. All of these facts show that hormonal distinctions between the sexes are not as great as we usually think. Referring to the hormones as either male or female constructs an absolute division that misrepresents the biology of the sexes.

Other hormones come into the picture, too. For example, androstenedione is another so-called male hormone, or androgen, but it tends to be present at higher levels in women than in men. Biology is never as simple as some would have us believe. There are more sex hormones than the four mentioned so far. They are all steroid hormones, meaning that they all have the same basic molecular structure, but they have different and interacting effects on the development of the genitals and other parts of the anatomy. They also have different effects on the brain. One such hormone, 5α-dihydrotestosterone, causes the genitals to develop into the male form (that is, it leads to growth of the penis and to the testes descending into the scrotum).

The body produces 5α-dihydrotestosterone from testosterone by the action of an enzyme called a reductase. Without this enzyme, and therefore without 5α-dihydrotestosterone, the genitals of genetic males look like those of a female. There is one family in Dallas, Texas, and another one in the Dominican Republic, who have a genetic condition that makes the males unable to produce the reductase enzyme that converts testosterone to 5α-dihydrotestosterone until they reach puberty.[8] These males therefore have a female physique until they reach puberty, at which point they appear to change sex. The penis begins to grow and the testes descend. Until then, these genetic males look like normal girls and are raised as such. At puberty they change to living as men.

Testosterone can also be converted to estrogen. In fact, most of testosterone's effects inside the cells of the brain depend on its conversion to estrogen. This means that the so-called male hormone has to be changed to a so-called female hormone before it can have an effect in the brain. The concept of male and female sex hormones is therefore most inappropriate when applied to the brain. Even though some of the various sex hormones may influence the development of parts of the brain, the concept of a female and a male brain is unlikely to be an accurate description of brain function. In making more of the differences than we do of the similarities, we influence how we think about biology and the brain.

The Environmental Candidates

From the moment a child is born, it is assigned to the male or the female sex and, in almost all cultures, is then treated as either a girl or as a boy. This decision has many ramifications for the parents, in the past relating to inheritance of the family's wealth, and these days relating to career choice and social success, carrying on the family name, and so on. Willingly or unwillingly, we teach female and male children different things. We can speak of the "environment" of a girl being different from that of a boy. The differences in the social environments surrounding girls and boys are fashioned by humans.

It has been argued that the different environments are not simply imposed on girls and boys by the society around them, but that the child seeks to put herself or himself in an environment typical for a girl or a boy. Girls are said to surround themselves by different environments from boys and seek to learn different things from boys because their genetic program drives them to do so. Having XX or XY genes is said to shape attention to certain things and ensure that the child will develop interests typical for his or her genetic (and hormonal) sex. Although this idea acknowledges the role of society and learning in determining sex differences in behavior, it also subsumes it within a framework of genetic determinism, encapsulating the environment within the genetic program for development. Genetic differences between the sexes are therefore seen as the primary, if not the supreme, factor guiding even those aspects of development that require learning. Any environmental influences on development are considered to be ultimately controlled by the genes.

At the end of chapter 1, we looked at development in terms of mutual interactions between the genetic program and experience (or learning), occurring at all stages of development and always subject to further change. In this form of mutual interaction, genes have a role, but it is neither central nor primary. This view of development therefore differs completely from the one outlined above, which sees genes as controlling what learning will occur by channeling the individual's attention to certain things and by constraining what experiences the individual will encounter and remember. It may be true, though, that some behavioral characteristics are the result of genetic influences that direct a person's choice of his or her social environment and so predispose the development of those behavioral characteristics. Matt McGue and Thomas Bouchard of the Institute of Human Genetics at the University of Minnesota have presented evidence indicating that this might be the case for some mental abilities and personality traits.[9] But it does not appear to be the case for sex differences in behavior.

If this idea of development were true in the case of sex differences, girls and boys would be expected to seek out different things to learn, even when given absolutely equal opportunity for education and equal encouragement to perform well on the same tasks. This would mean that sex differences in thinking and behavior would remain even if both sexes were given exactly the same opportunities to develop their abilities and interests. There is evidence that this is not what happens. Alan Feingold of Yale University examined sex differences in the performance in scholastic aptitude tests of children in the United States from 1947 to 1983. The usual sex differences were found—girls did better than boys in verbal abilities, and boys better than girls in spatial and mathematical tasks—but these differences, to quote the researchers, "declined precipitously" over the years surveyed.[10] That is, the recorded sex differences were greater in the early studies than in the later ones. Another study by Janet Hyde and colleagues at the University of Wisconsin[11] found the same result as Feingold. They reviewed a hundred published reports of studies on sex differences (they used the term "gender differences") in mathematics and compared the size of the difference in studies published before 1974 with that found in studies done after that date. The sex difference after 1974 was half what it was before. This is probably due to the changing attitudes about which careers are more appropriate for girls and which are more appropriate for boys.

It is therefore unlikely that sex differences in spatial, mathematical, and verbal performance are inherited and in this way immutably built into the biology of girls and boys. Instead, these differences are manifestations of social values held at a particular time. Boys and girls are not driven by their genes to select particular learning environments and to develop particular abilities. If girls and boys are given the opportunity of equal access to all forms of education, and the expectations for them to perform in sex-typical ways are reduced, most boys do not seek learning situations that will allow them to develop superior mathematical and spatial abilities, and most girls do not seek learning situations that will lead to superior verbal abilities. The social environment, not the genes, has a direct effect on the development and expression of at least some of the characteristics typical of girls or boys.

There may be other types of sex differences that are not so strongly affected by the changing attitudes of society. For instance, the sex difference in one other kind of spatial ability—the ability to match figures rotated at different angles—has remained unchanged during the past twenty years.[12] This might be because spatial ability really is genetically determined, as many scientists believe,[13] or because the changes that have occurred in society's attitudes to women and men have been insufficient to affect this particular trait.

Our physical appearance has a large effect on how other people treat us throughout life. Many studies have shown that people who are seen as physically beautiful in a particular culture have more success than average in personal relationships and employment. We can also decide how we will present ourselves, and this too determines how others treat us. Biological factors often have less to do with our actual physique than the cultural aspects of how we present ourselves. Transsexuals (or transgendered people as they are sometimes called) provide striking evidence for this. Many transsexuals portray themselves as the opposite of their biological (genetic and hormonal) sex and do so very successfully even before they undergo a sex-change operation.

Physical appearance as an external aspect of the individual—and this is sometimes at odds with genetic and hormonal sex—affects the social environment that surrounds the individual. By wearing clothes, we conceal our genitals from others, at least to a large extent. This means that we decide, either consciously or subconsciously, that another person is a woman or a man on the basis of general aspects of appearance that may have little to do with that person's actual genetic sex. We then react to that person differently according to whether we perceive that person to be a woman or a man. Transsexuals and transvestites tell us they are treated quite differently when they are dressed as a woman or as a man. In other words, we superimpose our values about sex, gender, and differences on other people depending on what sex we think they are. In turn, the social environment determines what aspects of behavior are acceptable for that person to express and what mental abilities they are allowed to develop.

Reducing the Complex to the Simple

The biological influences on the development of gender seem to be more tangible to scientists than the environmental influences. The potential biological causes of sex differences in mental processes can, superficially at least, be pinpointed to either genes or hormones. The biological differences between the sexes appear to be measurable, and they seem to be a simple solution to understanding the causes of sex or gender differences in behavior, even though the influences of genes and hormones are not so simple when we look at them in more detail. The term *environment* is often used as a collective term to refer loosely to any potential influences outside the organism in question. Biologists often use the term as if it were a single variable equivalent to a single gene or hormone, but it is a nebulous concept. As far as sex differences are concerned, the environment might embrace influences by other people, social systems, the media, nutrition, and so on.

Many geneticists ignore the influences of the environment altogether and turn to genes to provide all the answers, and endocrinologists tend to rely on hormones. Often without even investigating the potential effects of the environment on the differences in behavior that develop between women and men, these scientists make definite statements claiming that genes or hormones are the sole determinants of the differences. For example, in an article published in *Scientific American*, Doreen Kimura states: "Cognitive variations between the sexes reflect differing hormonal influences on brain development."[14] Despite the certainty with which this particular claim is phrased, there is no strong evidence on which to base it. The article describes some sex differences in the organization of the brain for processing and producing speech, and discusses hormonal differences between the sexes. But it provides no evidence that the hormonal differences between the sexes actually cause the differences in brain organization and function.

The whole area of sex differences—their existence and their causes—must bridge the gap between biology and psychology, between the more tangible elements of the brain and its chemistry and the far less tangible manifestations of its function in behavior. Human behavior is extremely complex and can be described and analyzed at different levels, in social terms and in terms of the individual's psychology. There are also physiological descriptions of some aspects of behavior, as well as molecular and physical ones. For example, consider someone driving a car. We might begin by describing the social environment that influences the kind of car that is preferred and the speed at which it would be acceptable to drive it. We could then describe the individual's psychological attitudes to driving and how they might influence the speed at which he or she will drive. Next we could describe the physiological processes in the body of the driver (such as the rush of adrenaline when the driver goes faster than the speed limit or the activity of the muscles controlling the foot on the accelerator). Yet a lower level of explanation would take us to describing the electrical activity in the driver's brain and then the molecular processes occurring in the brain.

It is generally accepted that there is a hierarchy of orders of analysis, each level corresponding to the traditional disciplines of science: physics, chemistry, physiology, psychology, and sociology.[15] This order is one of increasing complexity. So we might discuss sex (or gender) differences in behavior at the sociological level as an issue for society, or we may choose to discuss it at the psychological level and deal with comparison of the behavior of individuals. Or we might look at the physiological aspects of sex differences in behavior and even consider them at the molecular or submolecular level (for example, hormones and genes). All these levels of analysis can exist in parallel, provided that they do not contradict each other.

Behavior involves the interaction of the whole individual with the environment and with other individuals at higher levels of complexity than that at which genes and hormones operate. Although there is no reason why we should not discuss behavior in terms of genes or hormones, for example, such explanations are not complete in themselves, and much simplification is necessary. In fact, to equate social events with physiological events requires distortions, and we lose sight of many of the factors that may influence behavior. To do so reduces the explanation of behavior to a lower level of analysis. This form of thinking is referred to as *reductionism*.

To give an example of reductionism, depression in women is often said to be caused by a hormonal imbalance, if it occurs just before menstruation, after childbirth, or after the menopause. Psychosocial factors that might be equally or even more important are not taken into account, as the complexities of depression are reduced to the action of one or more hormones. They are often further reduced to the action of certain molecules within the brain. In this case, the depression is said to be caused by an imbalance of one of the neurotransmitters, either noradrenaline or serotonin, inside the brain. There are several different types of depression, and the symptoms of some forms can be reduced by treatment with antidepressant drugs. Some of these drugs (the tricyclic antidepressants and the monoamine oxidase inhibitors) act by altering the activity of noradrenaline and serotonin in the brain, but this does not prove that the depression is caused by an imbalance of these chemicals in the brain. The drugs may be having their effects by acting in ways that have nothing to do with the actual cause of the depression. It follows, then, that it would be simplistic to describe depression only in terms of noradrenaline and serotonin, although that is often done in medical textbooks and many research papers in pharmacology.

Reductionists believe that it would ultimately be possible to explain every aspect of behavior in terms of atoms and molecules in the brain. They assume that eventually the lower levels of explanation will subsume all the more complex levels of explanation. With this approach, they draw singular (or unitary) explanations for complex behavior, and they see explanations based on biological causes as more probable and superior to those based on social causes. We can see why such an approach is tempting: it is easier to measure the level of a hormone or a neurotransmitter, for example, and to control for it experimentally than it is to measure and control for a complex set of social events.

Although there are serious problems with the reductionist approach, it is the most common way of thinking in Western science, and it is taking hold widely in society. The very simplicity of reductionist explanations makes them attractive, but that does not mean that they are correct, especially when applied to behavior. To tackle a scientific problem with a reductionist approach can sometimes be useful in the initial stages of understanding, even in the case of knowing what causes behavior, because it helps us to design experiments to find out what is going on. But then the challenge is to put all the little pieces of evidence back together again. Often the whole picture is more than the sum of the parts. As in the nursery rhyme, even with "all the king's horses and all the king's men" it may be impossible "to put Humpty-Dumpty together again." Most reductionists forget that there is even a need to put the pieces back together again and beguile their listeners by claiming to explain the whole by only one small part.

Sociobiology

Sociobiology is a mode of thinking that is based entirely on reductionism. It postulates that there are genes controlling many different aspects of complex human behavior, including aggression, intelligence, homosexuality, altruism, selfishness, and sex differences in behavior.[16] Complex behavior is reduced to the functioning of the genes. Written in the mid-1970s, Richard Dawkins's *The Selfish Gene* has achieved wide popularity. To Dawkins, of Oxford University, genes are "selfish" in that they are seen as the driving force for reproduction, which ensures that they can replicate (multiply) and be passed on to the next generation. According to this line of thought, all the complex behavior of humans can be seen as a by-product of the need for genes to replicate. Sociological and psychological explanations of human behavior are subsumed by lower-order genetic explanations.

Most sociobiologists assert that social divisions between women and men are determined by the genes and have little, if anything, to do with social and cultural values.[17] The sex differences of the body and mind are said to have evolved to suit the demands of our distant past, when humans were hunter-gatherers, with women bearing and rearing children, and men competing for women and hunting to provide meat for the group. Evolutionary processes, it is believed, have selected genes that make women more nurturing and men more aggressive. These genes are said to still be part of our inheritance today. Humans in modern society are said to be driven by the same biological forces that might have been adaptive, or helpful, in our distant ancestors, although perhaps they are no longer adaptive today.

Sociobiologists also say that genes are the reason for sex differences in sexual behavior. It is said to be natural for men to be promiscuous and women to be monogamous because women put more "biological investment" into bearing and raising the next generation than men.

Women, it is reasoned, are more likely to be monogamous than men because their biological contribution to the next generation is greater than that of men, whose contribution to reproduction is to produce large numbers of sperm at little biological cost. Women produce eggs that are more rare (and costly to produce), and they must bear and nurse their offspring. This greater biological investment made by a woman means that she should procure and hold onto her partner so she can gain help in raising the children into whom she has put so much biological effort. Men, on the other hand, are driven by the need for their genes to multiply, and the best way to achieve that is to copulate with as many women as possible and have as many children as possible. For these reasons, sociobiologists argue, genes influence men to be polygamous and women to be monogamous, so promiscuity is seen to be biologically natural for men and unnatural for women. Biological imperatives are often used in this way to explain current social practices, as shown by the following statements by the sociobiologist Matt Ridley:[18] "Most people live in monogamous societies, but this may only tell us what democracy usually prescribes, not what human nature seeks," and "males are generally seducers and females the seduced. Humanity shares this profile of ardent, polygamist males and coy, faithful females with about ninety-nine percent of all animal species including our closest relative, the apes." Even this statement about apes is incorrect: sexual encounters among orangutans are, for example, much more varied than Ridley's simple model would imply.[19]

This biological justification for the current differences in sexual behavior and commitments to marriage and other forms of sexual bonding has met with much popular support and has been taken up by a new field of psychology, known as "evolutionary psychology." Evolutionary psychology was formed by combining sociobiology and psychology with a view to understanding the ills of modern society and how our minds work. It sees a mismatch between the modern environment and social changes toward equality of the sexes on one hand, and the ancestral environment for which our genes were adapted, on the other.[20]

It is interesting that sociobiology took shape during the 1970s, when women and minority groups were demanding change toward more equality in society. Sociobiology was part of a backlash to the changes taking place. Although the claims of sociobiologists were said to be new ideas at that time, it is not difficult to trace similar thinking to the previous century. The idea of the genetic determinism of human nature was accepted almost universally by the start of the twentieth century.[21] We seem to have arrived at a similar place in collective thought as we start the twenty-first century.

Put forward as new scientific discoveries in the 1960s and 1970s, the claims of genetic causes for human behavior were adopted by politicians and political groups and used as instruments to justify social policies maintaining inequality, even though they were mainly ideas with no substantial scientific evidence to support them. The idea of equality for women and men was alleged to be against nature, and this view was espoused at the same time that women's groups across the world were demanding equality. Since then, genetic explanations of human behavior have become increasingly accepted even though no more scientific evidence has been found to support them. Hand in hand with this trend has been an increasing acceptance of theories about genetic influences on behavioral development and of new scientific techniques that might, superficially, appear to confirm the primacy of genetic determinism.[22]

Steven Pinker, of the Massachusetts Institute of Technology, is one of the best-known evolutionary psychologists. He has taken up Dawkins's ideas, as well as those of other sociobiologists, and applied them to how the mind works. His position is clear from the following statement from his book *How the Mind Works*: "The ultimate goal that the mind is designed to attain is maximizing the number of genes that created it."[23]

This statement implies that the mind is a by-product of the genes and their biological imperative to replicate and be passed on to the next generation. As Pinker also says: "Even our bodies, our selves, are not the ultimate beneficiary of our design. . . . The criterion by which genes get selected is the quality of the bodies they build, but it is the genes making it into the next generation, not the perishable bodies, that are selected to live and fight another day" (43).

According to Pinker, the genes are the benefactors of our design.

They are therefore seen as the source of human behavior, including sex differences in monogamy and polygamy, aggression and the perception of beauty. This is clearly a reductionist's position. Of course, genes have a part in the development of sex differences and other behaviors, but it need not be any more important that other influences from both inside and outside the body, and the part played by genes might not be separable from these other influences.

Genes and Gender

On the subject of sex differences in behavior, one of the founders of sociobiology, E. O. Wilson, of Harvard University, said the genetic bias is intense enough to cause a substantial division of labor (between women and men) even in the most free and egalitarian of future societies.[24] In his opinion, even if the sexes receive identical education and equal access to all professions, men are likely to have a disproportionate role in political life, business, and science. By implication, women will continue to have a disproportionately greater role in child-minding, and maybe in professions providing nursing care and other kinds of nurturing. Sociobiologists think that women are genetically programmed for maternal feelings, for superior verbal abilities, and for performing repetitive tasks not requiring deep thought processes, whereas the genetic contribution of men prepares them for analytical thought and endows them with superior spatial abilities.[25]

Genes on the X and Y chromosomes are believed to direct the development of the brain, probably not without the intervention of the sex hormones, as we will see in chapter 3, and so sex differences in brain function emerge. Aggressive behavior has been seen as a male characteristic and, therefore, controlled by genes on the Y chromosome. This assumption alone has vast social implications, not least of which is the social acceptance of aggression in men. It has even been used to justify men's rape of women. Ridley suggests that violence may have been a route to sexual success in our distant ancestors, especially in times of turmoil, and that may be why rape is common in war even today.[18] Pinker has taken this idea a step further by suggesting that men banded together to fight because a conquest in tribal war allowed the victors to rape the women and so pass on their genes. Women do not band together and raid neighboring villages because, he believes, they do not have to compete for reproductive success. Instead, women may encourage men to form groups and fight. Having said this, Pinker points out that, if genes cause men to rape in modern society, that does not mean that rapists should not be held responsible for their actions. He believes we should pursue what he refers to as "the truth" in ivory towers undistracted by moral or political thoughts.[23] In my opinion, it is a misconception to think that any science of sex differences including rape could or should be pursued in the hallowed ivory tower without a thorough understanding of the broader social context.

Sociobiologists see no problem in accepting genetic hypotheses about the cause of sex differences in behavior. They see these differences as arising over countless generations by the process of natural selection as humans evolved from animals.[26] Authors Anne Moir and David Jessel say that these past two hundred years of industrialized society are a mere blip in our evolutionary development.[27] They are right about evolutionary time: humans evolved some five million years ago and this has been ample time for genetic selection to occur. But if the sexes were divided according to the labor of hunting and gathering, and that is debatable, this simple division did not persist unchanged up to the industrial revolution, as Moir and Jessel wish us to believe. Also, the mere availability of time during which evolution could have happened is not proof that it did happen.

Whereas Moir and Jessel present an extreme and popular view of sociobiology, sociobiologists are more circumspect about the details of genetic selection for sex differences in behavior, but generally believe that genes were selected for men to be hunters and women to be gatherers. These are ideas only, not proven fact. Pinker and other evolutionary psychologists, including John Tooby and Leda Cosmides, of the University of California, Santa Barbara, argue that hunting has been a major force in the evolution of the human mind.[23] It required planning ahead, an ability to know where to look for prey (spatial ability), and group communication. These attributes supposedly led to the selection of genes that improved the mental powers of our ancestors. Hunting is also said to have been exclusively an activity of men. In this way, mental superiority has been linked to male activities.

Sociobiologists consider that the superior spatial abilities in men have a genetic basis and arose over evolutionary time because males had to defend their territory or had to hunt. An ability to form mental maps of where things are and how they might change with the time of day or the seasons would have helped in both of these behaviors. To find food, it is said, males had to travel further than females from the home base. To hunt effectively, they may have required skills of throwing accuracy, both of which might have led to the evolution of better spatial skills in men than women.[28] Alternatively, male wanderings might have led to more success in reproduction. The hypothesis of male range size, as it is known, was considered in the context of animals and states that males increased the area over which they roamed so they could encounter more females and thereby increase their mating opportunities.[29] In turn, according to the hypothesis, this might have led to increased spatial ability and enlargement of part of the brain called the hippocampus, in which spatial information is processed. There is evidence in voles and kangaroo rats that males do have larger ranges, better spatial ability, and a larger hippocampus than females.[30] Although these findings have been extrapolated to humans,[28] this is far from justified beyond toying with the idea as fanciful speculation.

The converse argument can also be made: that women required better spatial abilities than men because they foraged for food that was growing, widely dispersed, and not moving.[31] This form of search would be enhanced by an excellent spatial memory of the location of objects placed in arrays, or patterns, and there is evidence that women perform better than men on tasks designed to test this type of spatial ability.[32] This shows that spatial ability is not a singular concept, but can be expressed in different ways. Recognizing this fact is important for refuting any simplistic claims of an association between male sex and spatial ability. It is just as important to consider the potential role of learning different forms of spatial ability that might occur quite subtly during early life. The abilities in which men and women differ in adulthood might be caused by permitting, persuading, or even forcing girls and boys to take part in different activities. In addition, there is no convincing evidence that, in our ancestral past, all men were hunters and all women were gatherers. Roles may or may not have been strictly divided according to sex.

All the claims about genetic causation of sex differences in behavior are universals, considered to be true for all peoples in different cultures. Because the genetic differences between women and men occur consistently in all human groups (XX and XY), sex differences should be found in all cultures and all ethnic groups, according to the sociobiologists' claims, unless that culture has some special sort of learning that suppresses the differences. It follows that even one exception to the rule can be taken as an indication that social factors might outweigh genetic causes. In very few cases have sex differences in behavior been examined in different cultural groups, but spatial ability has been tested in several different ethnic groups. In most cases, using the standard tests for spatial ability, men perform better than women, but there is an exception: in Eskimo culture, women have better spatial abilities than men.[33] This exception tells us that the sex difference in spatial ability is not a biological universal and that cultural factors might even reverse the direction of the difference.

According to the sociobiologists, selection for higher levels of aggression also results from the behavioral demands on ancestral males. Males made war and competed among each other, whereas females were homemakers, caring for the children and engaging in verbal exchange. These designated roles, it is said, account for the evolution of the traits that still characterize women and men in modern society. In other words, nature has dictated our differences over the long course of evolution. This is an example of a process called sexual selection. Again, Moir and Jessel state the sociobiologists' position in a nutshell: "For most of our past, we lived in communities which depended for their very survival on hunting animals and gathering plant food. Men, with their greater strength and stamina, their roving tendency, and their greater skill in relating the spear to the space occupied by the prey, did the hunting, an unpredictable and dangerous activity. Women gathered nuts, grain, and grubs—a safer and surer pursuit."[27]

This statement is not substantiated by anthropological evidence, nor is there evidence that any form of sexual selection has continued to operate at the genetic level across generations of humans up to the present day. Learned patterns of behavior can persist over generations just as readily as patterns programmed into the genes.

Sociobiologists are not merely speculating about sex differences and genes. In so doing, they are also constructing a framework of ideas about what is natural and what is not. Women who enter professions that are typical of men are therefore seen as unnatural and going against their biology; so too are men who take up professions using abilities considered to be typical of women. These "unnatural" women and men are considered to threaten the fabric of society, as seen and maintained by those (scientists, politicians, business leaders, and the general public) who see genes as paramount in causing sex differences in behavior. The notion that genes cause sex differences has more to do with social attitudes than scientific proof.

More recently, genetic ideas about sexual selection have been applied to physical attractiveness and mating success. It has been postulated that symmetry of physical features (not only of the face but also many other parts of the body) leads to more mating success than asymmetry.[34] The left and right sides of the face and other parts of the body are asymmetrical to varying degrees, with some individuals being more asymmetrical than others. This is called "fluctuating asymmetry," and it is found in all species. But Randy Thornhill and colleagues, of the University of New Mexico, have claimed that symmetry is the ticket for beauty and sexual success.[35] Human faces that are more symmetrical in appearance tend to be seen as more attractive and so more likely to attract a sexual partner. The reason for this, it is claimed, lies in that fact that symmetry represents health and hence good reproductive potential. Asymmetries may result if a person has "bad genes," carrying mutations, or has been attacked by a pathogen or a toxic substance during development. The same selection for symmetry is said to happen in a wide range of species, as well as humans. Indeed, Thornhill formulated his theory when studying the mating behavior of the Japanese scorpion fly. He found that males with the most symmetrical wings had more mates than average.

The same idea was applied to other species, and some evidence in support of it has come to light, apparently even in humans. Symmetrical humans, according to Thornhill and colleagues, have more orgasms than asymmetrical ones.[35] The evidence was obtained by selecting eighty-six heterosexual couples from among students studying a course in introductory psychology, and measuring the widths of the men's feet, ankles, wrists, and other parts of the body, as well as the sizes of their ears. The degree of left- and right-side asymmetry for each of these features was calculated, and these values were correlated with the frequency of sexual orgasm reported by their partners. This was determined by getting the subjects to answer a questionnaire about their sexual behavior. Women married to the more symmetrical males claimed to have more orgasms during intercourse than women married to less symmetrical males. The researchers interpreted this result as support for the idea of sexual selection for symmetry in humans, as in flies and other species. That is, they believe that the correlation of symmetry and reproductive success is a basic principle of biology and that it has something to do with selecting the best genes for the next generation. We can only guess what that something might be. There is no known mechanism that links human genetic fitness to symmetry or fluctuating asymmetry, nor is there any known connection between female orgasm and increased likelihood of reproductive success, despite the fact that Thornhill and colleagues argue that it aids in sperm retention and so may assist conception.

But even if the findings of Thornhill and colleagues apply generally to humans (in different ethnic groups, different socioeconomic classes, and so on), there are different ways of interpreting the data. Self-reporting of number of orgasms is notoriously unreliable; perhaps the partners of symmetrical men are prone to exaggerate. Even if symmetry of the male partner does correlate with frequency of female orgasm, this does not tell us anything about the factors causing this relationship. The causes might be various and complex. One could make many speculations, but the fact remains that the scientific basis on which all of these speculations are based is very shaky indeed. It is tempting to see this research as an attempt by biologists to get into the field of sociology with no knowledge of the social forces impinging on the behavior under consideration.

Genetic explanations for sex differences in behavior are part of a broader wave of opinion denying the importance of social forces in the development of human behavior. The genetic code is increasingly being seen as the key to all our behavior.

3

Gay Genes?

Do genes on the X and Y chromosomes influence brain development such that a person thinks and behaves in either feminine or masculine ways? It has long been assumed that they do. Genes that might affect brain development and function have been thrust into the limelight recently by the availability of new techniques to map the location of genes on the human chromosomes, and exciting new discoveries are being made about the genetic makeup of humans. However, there is a negative aspect of mapping genes. The focus on genes has reinforced the long-standing notion that human behavior has simple genetic causes. We saw in chapter 2 that sociobiologists discuss the causation of behavior in this way, and now the availability of new molecular techniques for studying the genetic code has lent a spurious respectability to their views. This has happened despite the fact that gene-mapping technology has had little, if any, success so far in linking behavioral characteristics to particular genes in humans.

People often speak of genes as if they hold the program for an entire individual, governing even how the individual interacts with other individuals and other aspects of the environment. We commonly hear about genes controlling aggression, intelligence, schizophrenia, alcoholism, homosexuality, and so on. It is as if a whole range of complex behaviors is read off from the genes, like a recipe from a book. The explanation of complex human behaviors has been reduced to the level of the genes. This is reductionist thinking, as it collapses complex behavior (at a higher level) into explanation at the molecular (lower) level of the genes. As Steven Rose of the Open University in the U.K. has explained,[1] reductionist explanations have their place in the science of living systems but they are not primary, and they become "crudely ideological" when applied to human behavior.

The expressed aim of the Human Genome Project is to map the location on the chromosomes of all the human genes. Many of these genes code for physical features and aspects of a person's physiology. There are clear cases where single genes influence features, such as eye color and hair color; these features depend on the gene making a single protein that influences one of these traits. Some genes code for proteins that are enzymes, some of which are important for the functioning of the brain. If a person has a genetic defect that prevents the body from making one of these enzymes, brain function can be affected in many ways. For example, phenylketonuria is a genetic condition in which the enzyme that converts the amino acid phenylalanine to tyrosine cannot be made, resulting in mental retardation. It is possible to find out where on its chromosome the gene responsible for making this enzyme is located. There are other examples of genetic defects that affect behavior in very general ways, but it is not so easy, or even possible, to find out whether a single gene affects smaller units of behavior.

It is far too simplistic to think that the mapping process can be extended to all types of human behavior because it is most unlikely that any particular behavior pattern depends largely on the action of a single gene, or even a string of genes. Although genes may play a part somewhere along the line, their input is usually so remote and so malleable, depending on interactions with the environment, that it would be impossible to make any direct association of the behavior with a gene or genes. Nevertheless, many molecular geneticists see mapping the locations of genes controlling complex behaviors as a possibility, and talk as if they are just about to find the gene or genes controlling a number of human behaviors. They believe that, in the human genome, they have found "the book of life" from which they might read all the characteristics that will be, or might be, expressed by an individual. Some molecular geneticists refer to the genetic map as the "code of codes," as if they are about to unravel the entire secret of life.[2] This is the fashion of the moment, and it has turned our eyes away from those dimensions of the individual that depend on experience. If influences from the environment are taken into account, they are seen to be constrained by the genetic code, meaning that genes determine even the environment a person "chooses" to be in.

Molecular geneticists have gone beyond linking genes to physical traits of individuals and are now suggesting that genes control all manner of complex behaviors. Some have even declared that the nature-nurture debate (that is, between genes and experience) is now over, as genes have supposedly been shown to control human behavior.[3] They believe that we are on the verge of eliminating many problems of society, such as drug abuse, homosexuality, and even homelessness. In its extreme form, this way of thinking leads to a new form of eugenics (planned reproduction) to eliminate these aspects of society. The aim is to identify the genes responsible for these behaviors and then, apparently, to abort fetuses that are found to have them or advise individuals who have them not to reproduce. Another alternative for the future may be gene therapy.

It is, of course, a crude interpretation of genetics to suggest that all individuals who might be carrying, say, a hypothetical gene "for" homosexuality or lesbianism might express that particular behavior. Many would argue that to carry the gene means only that there is a propensity or predisposition for the behavior; that is, the gene may be expressed in certain environments but not in others. So, if there is a gene for lesbianism, it might only increase the chances of individuals carrying it being lesbian. It follows that the effects of this gene might be prevented in certain environments. If these environmental conditions were known, it is argued, they might be avoided, and the "problem" could be eliminated. In this case, we are not talking about using eugenics to eliminate the problem, but using social control by manipulating the environment (family and school life, gender identity, learning, and so on) so that the offending gene cannot be expressed. Alternatively, if individuals carrying the gene or genes can be detected, certain professions (such as the army) or business companies might use the information as a basis for excluding homosexuals and lesbians. These practices would be as heinous as eugenics.

Gene Mapping and "Normality"

The behaviors that molecular geneticists have so far attempted to link to genes and to locate on the chromosomes (known as mapping onto the genome) are those considered to lie outside the bounds of normality. In that sense, behaviors that might be typical of females or males have so far escaped scrutiny, but this is not so for homosexuality. Homosexuality is usually referred to as a "sexual orientation" in the titles of scientific publications, which implies an abnormal sexual preference because it is not a heterosexual orientation that is examined. It has become common for people to speak of "gay genes," or a single "gay gene," and even a "gay brain," as if the genes themselves, or the brain itself, were somehow gay. It is as if the genetic material holds the blueprint for "gayness" and makes its stamp on the brain. Being gay therefore becomes something genetically abnormal, a flaw in the plan. There is no attempt to find the genes for heterosexuality rather than homosexuality. The search for genes more often than not begins with defining a character as abnormal. Socially, this is a poor starting point; it is even worse for a scientific investigation.

Deciding what is normal depends on choosing one type of behavior, measuring it, and then assuming that the entire human population can be described according to that single dimension. In a mathematical or statistical sense, normality is often represented by plotting a graph.[2] We might take height as a simple example. The height of all individuals in a population can be plotted in a graph (or histogram), with height along the horizontal axis and the number of individuals at each height plotted on the vertical axis. A bell-shaped curve is generated: most individuals are clustered around the average height for the population and a few diverge greatly from that by being either much shorter or taller than average. There will always be a spread of individuals diverging from the mean, but most will not be very much shorter or taller than the mean. This is what we mean by statistical normality.[4] This form of normality has nothing in particular to do with normality in a social sense. The latter is judgmental and based on religious or social beliefs and has nothing to do with any bell-shaped curve.

It is reasonable to plot a physical feature such as height as a distribution of scores to demonstrate statistical normality, but problems arise when the same unidimensional approach is applied to human behavior. For example, it is common to measure and plot IQ in this way, as if the intelligence of an individual could be encapsulated into a unitary score determined by a single test. There are many different forms of intelligence, and the score obtained from just one IQ test does not tap into all of these. The result of plotting the distribution of IQ scores for the population is a bell-shaped curve, like that described above for height. Yet this is more than a statistical exercise because society makes decisions about the education and future of individuals whose IQ scores fall on the extremes of the bell-shaped curve (that is, those considered to be "abnormal" in a statistical sense). Those with scores falling in the low range are excluded from normal schools, whereas those with scores in the high range are said to be gifted and are often allowed certain privileges in education.

Essentially the same thing applies to other behaviors plotted in the same way, masculinity versus femininity being another prime example. It is assumed that each person's behavior can be given as a single score on a scale ranking from extremely masculine to extremely feminine. Plotting the scores would lead to one bell-shaped curve for men and another for women. Men whose scores fall well outside the average for men, and women whose scores fall far outside the average for women, are said to be not just unusual but abnormal.

It is these behaviors perceived to fall outside the accepted norm that come under scrutiny. We do not exactly plot sexual orientation on a distribution curve, but we have in our minds a concept of sexual normality (also referred to as heterosexism) and a concept of what falls outside this range. This position of setting normality versus abnormality is the first dubious step in any scientific investigation of sexual preference. The next step is to discover what causes the perceived abnormality. This is why there is so much research to look for genes that might cause an individual to become homosexual. The motivation is to establish the genetic link and then to identify the gene or genes responsible so that individuals carrying these genes can be identified.

We must recognize from the beginning that this sort of scientific research is not neutral, but is done with a social application in mind. Finding a biological cause for difference sets the difference in stone. The person singled out will have to wear the label of being different (homosexual, say), and society will formulate fixed attitudes and laws accordingly.

Homosexuals have been studied from almost every aspect of biology imaginable. They are the interest of neuroanatomists, who measure the size of certain areas of the brain with the aim of finding the underlying cause for homosexuality, as well as molecular geneticists in search of the gay gene or genes. So what have they found?

First, it is important to mention that research on sexual orientation is never far removed from research on gender. Gay men and lesbians are considered to fall outside the norms for their particular sex, and homosexuality is often said to be associated with adopting the patterns of thinking and behavior characteristic of the opposite sex.[5] We can therefore expect the outcome of research on homosexuality to have an impact on the division between women and men. If sexual preference were found to have its roots in an individual's genes or hormones, there would be yet another buttress to hold up the dividing wall between women and men. Without doubt, the concept of the hypothetical gay gene or the gay mix of hormones would be extended beyond sexual preference to include those behaviors and ways of thinking associated with gender.

One of the main obstacles to research on homosexuality is finding appropriate subjects to take part in the study. As some surveys have shown, most individuals in a society have taken part in some form of homosexual activity at some time during their lives,[6] so the distinction between heterosexuals and homosexuals is not as clear-cut as one might imagine. Although homosexual behavior in many young people might be seen to be experimental and so different from genuine homosexuality, which persists into later life, this may also be said of heterosexual behavior. Homosexual experimentation may also take place at any age, particularly among women.

Researchers might confine the homosexual group in a study to self-professed individuals who specialize in homosexual relationships, but it should be remembered that the labels *homosexual* and *heterosexual* may make the differences more apparent than they are in practice. There is ample evidence that the lifetime of an individual may involve homosexual and heterosexual periods or even concurrent homosexual and heterosexual relationships. Labels of homosexuality and lesbianism are therefore more accurately applied to sexual practice rather than to individuals as statements of identity, although, for a core of exclusively homosexual people, homosexual identity is a strong and often political expression of practice.

Few studies searching for a "biological cause" of homosexuality bother to define what they mean by the term *homosexual* or give a full description of the volunteers who take part in the study. It goes without saying that individuals who volunteer to be subjects for scientific research in such studies are a particular subpopulation of those whom they are taken to represent, whether they are homosexual or heterosexual. Self-selection has biases that mean the sample is not random. For example, the subjects taking part in the study may be from a particular social club or social class. Rarely are the homosexual and heterosexual volunteers matched for class, lifestyle, personality, past experience, and so on. This means it can never be clear that the scientist is studying aspects of his or her subjects' sexual preference and not aspects of their personalities or past experiences that led them to volunteer for experimentation. In addition, it is not difficult to see that volunteering in these circumstances means something quite different for homosexuals, who know they are not accepted and are the focus of the research, and heterosexuals, who make up the so-called control group. These differing factors, whatever they might be, are as likely as sexual preference to be under the microscope, although they are usually unaccounted for, unnoticed, and ignored.

Gay Genes

A report published in *Science* in 1993 caused widespread debate in both the public media and the scientific literature. Dean Hamer and colleagues, of the National Institutes of Health, Bethesda, claimed to have isolated a gene sequence (that is, a string of genes on part of a chromosome) that might code for male homosexuality.[7] It was said to be located in a small stretch of DNA on the X chromosome. This piece of DNA, which they called Xq28, contains hundreds of genes.

The researchers recruited homosexual men into the study through an out-patient HIV clinic or through advertisements in "homophile publications." They began by interviewing the subjects about the incidence of homosexuality among their relatives. They selected forty pairs of homosexual brothers who had homosexual relatives on their mother's side of the family but not on their father's side. The reason for this choice was that the researchers were interested in looking for sex-linked gay genes carried on the X chromosome. Because a male inherits his Y chromosome from his father and his X chromosome from his mother, any characteristic carried on the X chromosome will be passed on from mother to son, not father to son. By selecting families in which homosexuality was more common on the mother's side of the family than on the father's, Hamer and colleagues were narrowing down their search for the gay gene(s) because they could concentrate on the X chromosome and exclude all the others. This is why they chose gay men with gay brothers, uncles, and cousins on their mother's side of the family. Gay men with gay uncles and cousins on the father's side of the family were excluded, as were pairs of brothers who had more than one lesbian relative, on the grounds that they were looking for a gay gene, not a lesbian gene.

Hamer and colleagues took samples of blood from the homosexual brothers and from other family members. These samples were used to amplify the DNA, and then they identified genetic markers, which allow detection of particular sequences of genes on the chromosome. The genetic markers on the X chromosomes from each individual were then matched up with those of each subject's homosexual relatives. The idea behind this genetic analysis is that, on average, each pair of brothers will have in common about half of their DNA on the X chromosome (and on all of their other chromosomes too) and, if both brothers are homosexual because they inherited the same gene, then the gay gene must be located on the parts of the X chromosome that they have in common.

Five markers at the tip of one arm of the X chromosome (the Xq28 region) were found to be matched in thirty-three of the pairs of brothers. The researchers therefore claimed to have isolated a gene sequence that gives a person a predisposition to become homosexual. They concluded by saying that their study needs to be repeated and that the Xq28 region of the X chromosome needs "fine mapping" to identify "a specific gene . . . where and when it is expressed and how it ultimately contributes to the development of both homosexual and heterosexual orientation."[8]

This statement shows that Hamer and colleagues believe that a single gay gene will be found in the Xq28 sequence, and that the tendency to be homosexual is somehow embodied within a single gene. This belief is, to my mind, not sensible. To reduce the complexities of an individual's choice of sexual partners to the operation of a single gene ignores the complexities involved in choosing a sexual partner, at least in most cases, and it ignores the plasticity of the brain in response to experience and learning. The idea of a single gene, or even a sequence of genes, coding for homosexuality denies historical and social influences on human sexual desire. It is a result of reductionist thinking and greatly simplifies the factors that interact to determine the choice of a sexual partner. Is the gay gene expected to exert its effects in the cells of the brain? Some researchers think so, as made clear by the following statement by Richard Pillard and Michael Bailey of the Boston School of Medicine: "The gene(s) in question is presumably expressed in the brain, but finding it and learning what it does will require considerable further effort."[9]

The study of Hamer and colleagues can also be criticized for statistical and technical reasons. Moreover, a recent attempt by George Rice, of the University of Western Ontario, and others has failed to find any genes for homosexuality in the Xq28 region of the X chromosome.[10]

There are also the common problems of subject selection and methods of interview. Details of the techniques used in the interviews were not mentioned. There were also some more concrete problems. For example, the researchers based their calculations on the assumption of a 2 percent prevalence of homosexuality in the general population, which is a low estimate. As science writer Bruce Bower pointed out, it is more generally accepted that the prevalence of male homosexuality is somewhere between 4 and 10 percent.[11] According to Evan Balaban of Harvard University and Anne Fausto-Sterling of Brown University, if the data are recalculated using the figure of 4 percent, then some of the claimed findings lose their statistical significance.[12] They have also expressed concern that DNA samples were taken from less than half the mothers in the linkage study, and the rest of the calculations were based on estimates. As the Xq28 sequence is on the tip of the X chromosome, this problem is heightened, as there are inaccuracies in tracking down markers on either side of a sequence in this location at the tail end of the chromosome.

Apart from these problems, even if the results are supported by further studies, it is possible that the Xq28 gene sequence specifies some behavior other than homosexuality. All the subjects were openly gay. This separates them from the very large population of gay men who prefer not to disclose their homosexuality for fear of reprisals. Many men conceal their homosexual activities because they are married, for example, or hold professional or other positions that might be jeopardized were they to be open about being gay. So, even if the results are supported by further studies, it is possible that the Xq28 sequence coded for some other aspect of behavior that characterized the volunteer group of gay men. Perhaps, as Hamer and colleagues also acknowledged,[7] the sequence contains some genes that might lead to a more extrovert personality, although I would be just as reluctant to argue for the existence of genes that influence such aspects of personality as I would for sexual orientation. This example is raised merely to demonstrate the complexities of interpreting a result, even if the experimental protocol did happen to be convincing.

When the study was published in 1993, the authors tried to encourage other research groups to repeat their work. In fact, the same research group extended their own study by reporting in 1995 on two new series of families containing either two homosexual brothers or two lesbian sisters.[13] The Xq28 sequence match was found for the families with gay men, in line with the previous result, but not for families with lesbians. This result suggested to the researchers that lesbian preference has nothing to do with the Xq28 gene sequence. Indeed, it would seem that different factors are involved in male homosexuality and lesbianism, irrespective of whether they are of a social or genetic nature.

In all their published papers, Hamer and colleagues have stressed that sexual orientation appears to be influenced by many factors. The homosexual men in the study were selected on the basis of particular family patterns of homosexuality. Across generations, the homosexuality had to be on the mother's side of the family (that is, a matrilineal line) and not the father's side. This meant that they could focus on the X chromosome, as this is passed on from one generation of males to the next by mothers, whereas fathers pass on to males only the Y chromosome. The Xq28 gene sequence did not feature in families without the particular matrilineal line of homosexuality, despite the fact that these families also contained gay men.

The group of men featured in the research of Hamer and colleagues represents only one small section of all homosexual men: those having a known or openly gay relative. Media reports of the research have invariably lost sight of this fact. Moreover, the researchers point out[11] that, even within the selected population that they studied, the Xq28 region was neither necessary nor sufficient for homosexuality to be expressed, meaning that there is no one-to-one relationship between being gay and having the Xq28 gene sequence.

Any research purporting to have found a genetic cause for homosexuality receives wide coverage in the scientific literature and in the general media because it has such important implications for society. The "Research News" section[14] of the issue of *Science* in which the Hamer group's original research was published in 1993 expressed excitement about the Xq28 discovery, asserting that the study was better controlled than previous ones claiming to have found genes for schizophrenia, alcoholism, and manic depression. The author of this article seemed to base most of his enthusiasm on the fact that Hamer is a good, solid scientist with a reputation for unraveling genetic codes, particularly in yeast. This might explain the reductionist approach: a biologist who has worked on genetic codes in yeast might not be expected to understand the social and psychological factors surrounding the topic of sexual orientation. The publication of the "Research News" article mentioned the fact that similar claims for locating genetic markers for schizophrenia, alcoholism, and manic depression had been short-lived. The author did not question the interpretation of the gay gene proposal at the social level of the study, nor did he venture to consider the complexities of development of brain and behavior. On the contrary, he seemed anxious to know how the proposed gene for homosexuality might work: "What protein does it code for?" This question implies a belief that all the complexities involved in choosing a sexual partner are manifest by the action of a single protein in the brain. In my opinion, this is most unlikely even if that supposed protein affects the level of a neurotransmitter in the brain. Sexual preference is not likely to depend on a single gene, a single neurotransmitter, or a single place in the brain.

Studies of Family Patterns and Twins

Studies of the human genome such as those just mentioned usually begin with an investigation into whether the behavior of interest runs in families, and especially whether it occurs in twins. If it is more frequent in certain families than in the general population, it is usually assumed that the behavior must be inherited. If true, this would mean that the families in which the behavior appears in each generation might have genes that influence it.[15]

Several studies have found that being gay or lesbian seems to run in some families.[16] At first this might seem to suggest that being gay or lesbian is passed on genetically, but actually this can occur without having anything to do with genetic inheritance. Families can have cultures that are more or less tolerant of homosexuality than society in general. Some families may even encourage it in ways that are not at all obvious, either to the researcher or to the family members themselves. Whatever the reason, as long as sexual attachment to the same sex is a common variation on human sexual behavior, as indicated by the Kinsey report[6] among others, it may be expressed or suppressed in some families but not others.

Social factors are also involved in establishing similar behavior patterns in twins. However, these are often ignored when twins are used in attempts to discover genetic causes. Twins have been the focus of many studies searching for a gene or genes that might determine sexual orientation. The approach has been to compare the frequency of homosexuality in pairs of identical and nonidentical twins.[17]

Identical twins, or monozygotic twins as they are also called, develop from the same fertilized egg, so they have the same DNA in the nucleus of each cell. Each pair of identical twins shares the same genes. Nonidentical twins, or dizygotic twins, develop from separate fertilized eggs and so have different sets of genes. They have, on average, only half of their genes in common. Identical twins are always the same sex, whereas nonidentical twins can be either the same sex or opposite sexes. In theory, a characteristic (physical or behavioral) that is similar in monozygotic twins and different, or not so similar, in dizygotic twins is considered to be inherited. But this line of reasoning denies the strong possibility that, because they look alike, identical twins are more likely than nonidentical twins to be treated similarly.

To allow for this possibility, it is better to study identical twins that have been separated at birth and raised apart. This way, shared behavioral characteristics are less likely to have resulted from the fact that the twins were raised in the same way. Nevertheless, being raised apart does not mean that all similarities in experience and learning are removed. The physical characteristics of a person (such as height or hair color) have a large effect on how he or she is treated by others. Because identical twins who are reared apart retain many similarities in appearance, at least some of the ways in which they are raised and reacted to by others might be the same despite their separation. As a result, both twins would be expected to develop some of the same patterns of behavior even though they were separated at birth. This would not deny a role for genes in the process—genetic influences would have determined the similarities in physical appearance—but it makes the possibility of special genes for behaving in particular ways less likely. Furthermore, it is unlikely that twins separated at birth are raised in entirely different environments because most adoption agencies strive to place twins in families of equivalent social and economic status, and rarely are they placed in families of different cultures or religions.

We should also not forget that twins develop in the same uterus and therefore, until birth, share the same environment. Little is known about the effect of the uterine environment on behavior in later life,

but we know it can affect the development of brain and behavior. In
the uterus, the fetus hears its mother's heartbeat and voice. After birth,
the newborn likes to listen to the mother's heartbeat and voice and will
suck a teat that does not deliver milk but allows it to hear these
sounds.[18] In other words, the baby will try to hear the sounds it heard
inside the uterus, whereas it will not try to hear an unfamiliar heartbeat or voice. It is therefore likely that sharing the same uterine environment before birth might affect the development of some aspects of
behavior. This means that separating twins at birth does not mean that
all experiences will be different.

Comparison of the similarities between monozygotic and dizygotic
twins controls for the effect of sharing the same uterine environment.
If the monozygotic twins are more similar than the dizygotic twins,
the shared uterine environment is unlikely to be the reason for this.
Common experience immediately after birth must also be important
in causing similarities later in life. Yet I have never seen a study of twins
separated "at birth" that states exactly when each pair of twins was separated. Even the first few hours after birth may have profound effects
that cause the same behavior patterns in twins.

Before the 1970s, much emphasis was placed on twin studies and
the inheritance of intelligence, measured in IQ tests. These studies fell
out of fashion when the work of their main protagonist, the British
psychologist Cyril Burt, then of the University of London, was dis-
credited by Leon Kamin, of Princeton University.[19] Kamin showed
that Burt had overstated his case for the genetic inheritance of IQ, and
that he had even invented much of his data. These considerations led
to a waning of interest in twin studies until the recent resurgence of
interest accompanying the development of new techniques for studying the human genome.

A study by Michael Bailey, of Northwestern University, and
Richard Pillard, of Boston University School of Medicine,[20] reported
in 1991, looked at the sexual orientation of brothers who were identical twins, nonidentical twins, or adoptive siblings, in which one of the
pair was identified as homosexual. The homosexual subjects "volun-
teered" for the study by replying to an advertisement. Each was questioned about the sexual orientation of his brother, and in some cases

this information was checked by questioning the brother directly. There were between fifty and sixty individuals in each group. The incidence of homosexuality in the brothers of the identical twins was 52 percent; in the brothers of nonidentical twins it was 22 percent; and in the adoptive brothers it was 11 percent. The same researchers have recently conducted a similar study of lesbians,[21] reporting that the incidence of lesbianism among the sisters of identical twins is 48 percent, of nonidentical twins 16 percent, and adoptive sisters 6 percent. In this case there were seventy-one identical twins, thirty-seven nonidentical twins, and thirty-five adoptive sisters, all of whom entered the study in response to advertisements placed in "lesbian-oriented" publications in the United States. Here one notes again the problem of using a self-selected sample. Again, the sexual orientation of the relatives was assessed by asking the volunteers and, where possible, checking the answers with the relatives. The results were interpreted as evidence for the genetic inheritance of gay and lesbian orientation because the incidence of both twins being either gay or lesbian was higher in the identical twins that in the other two groups. Yet half of the identical twins did not have the same sexual orientation; that is, one twin was heterosexual and one homosexual, despite the fact that both twins had identical genes and shared the same womb. If it is argued that genes determine whether someone is gay or lesbian, then the study could be viewed as showing that they do not do this to any great extent. The researchers ignored this fact.

Their conclusion that homosexuality is genetically inherited assumes, of course, that any of the unknown environmental influences on this behavior are acting equally in all three groups. Because these twins had not been separated at birth, it is reasonable to suggest that identical twins are treated more similarly than nonidentical twins or adoptive siblings, and that this could be at least part of the explanation for the greater incidence of both twins being gay or lesbian in the identical than in the nonidentical twins. Other scientists have criticized this study and similar ones for the same reason.[22] In other words, because the identical twins were raised together, similar environmental influences rather than genetic factors could have caused some of them to adopt the same sexual orientation. Although the authors consider this possibility, they discount it on the grounds that identical twins who are treated similarly in terms of being dressed alike have no greater similarity in IQ scores than those who are not treated similarly. But being dressed alike may not be an environmental factor relevant to either IQ or sexual orientation. As we have no idea what diverse environmental factors shape our sexual preferences, it is impossible to decide whether the greater common incidence of homosexuality in identical twins than in nonidentical twins is due to genetic causes or the effects of experience.

I know of only one study of lesbianism in twins who were reared apart. Although the study was limited to only four pairs of identical twins, none of the four pairs showed the same sexual preference.[23] This result indicates that genetic factors have no role in determining sexual orientation, but the sample size is too low to be sure of this. Yet, as a case study, it is not in line with the conclusion reached by Bailey and colleagues.

A more recent study sampling a very large number of twins failed to find any difference in the incidence of homosexuality in identical and nonidentical twins.[24] This points away from a genetic cause for homosexuality in the broader sense.

Measuring the Homosexual Brain

Simon LeVay, then of the Salk Institute for Biological Studies, San Diego, and Dean Hamer, mentioned previously, began an article published in *Scientific American* in the following way: "Two pieces of evidence, a structure within the human brain and a genetic link, point to a biological component of male homosexuality."[25] This is not a definite statement, but it links genes to brain structure and leans toward the idea that homosexuality is biologically determined. They also speculated that the protein encoded by the Xq28 region might act on the developing brain to affect the survival of nerve cells in one small region (referred to as $INAH_3$) in the lower part of the brain, known as the hypothalamus.

In the past decade, several groups of scientists have been involved in measuring the size of various regions in the brains of people after death. Comparisons have been made between men and women, and homosexuals and heterosexuals. The underlying theme is that genetic factors have caused the male brain to differ from the female brain and that this extends to differences between the brains of homosexuals and heterosexuals. A man is considered to be homosexual because he has a "gay brain," which in turn has come about because he has "gay genes."

Simon LeVay[26] has measured the sizes of four regions in the hypothalamus. Sampling the postmortem brains of gay men would normally be a very difficult task, but LeVay was able to obtain the material he needed from gay men who had died of AIDS. He found evidence that the average size of the $INAH_3$ region of the hypothalamus is smaller in homosexual men than in heterosexual men, and so is more like that of women (fig. 3.1). Although LeVay used a sample of only six women, all of whom he presumed to be heterosexual, it was already known that $INAH_3$ is smaller in the brains of women than men. In fact, this is why LeVay chose to look at this region.

Could such a small area of the lower region of the brain control the decisions involved in the choice of a sexual partner? Is that where the hypothetical gay gene is acting? Most of us on reflection would contemplate, or at least hope, that higher mental processes are involved. The choice of a sexual partner is more complex than we might at first think. It may involve taking into account personality, conversation style, the context of the meeting, and many other factors in addition to physical appearance. The parts of the brain involved in making these decisions are likely to be not just in the hypothalamus, but also in the cortex of the brain, where higher levels of thought are carried out. In studies using animals (mainly rats), a region of the hypothalamus called the medial preoptic area has been shown to regulate the level of sexual activity and to control some of the movements of the body used in sexual intercourse, particularly of sexual behaviors typical of males. But this region of the hypothalamus has not been linked to the choice of a sexual partner in animals, and it is even less likely to be the region of the brain where all of these decisions are made in humans. Yet that is what LeVay implies.

LeVay's experimental design had several flaws. All nineteen of the brains of homosexual men examined came from men who had died of

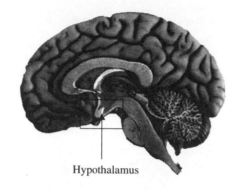

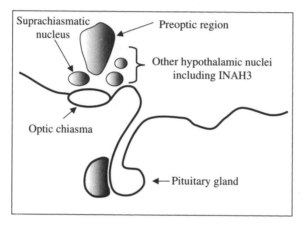

FIGURE 3.1 The hypothalamus is shown enlarged. As in figure 1.1, the front of the brain is to the left of the page and the back of the brain is to the right. Note the nuclei (collections of nerve cell bodies) in the hypothalamus. There are a number of other nuclei that have not been indicated.

AIDS, whereas less than half of the sixteen brains of heterosexual men were from AIDS patients. Because AIDS is frequently associated with the central nervous system being attacked by either the AIDS virus itself or other viruses that gain access to the brain because of the weakened immune system, this or any other as yet unknown aspects of AIDS could have caused the differences LeVay observed. The result may therefore have been caused by the AIDS infection and so have nothing directly to do with homosexual choice. Also, the sexual preference of the non-AIDS subjects was assumed to be heterosexual, rather than known to be so.

To counter the possibility that viral attack had diminished the size of $INAH_3$ in the homosexual group, LeVay argues that the $INAH_3$ area in the brains of the assumed heterosexual men who had died of AIDS was the same size area as that in their counterparts who had not died of AIDS.[21] But this comparison included only six brains from AIDS victims and ten brains from those who had not died of AIDS. Such a small sample size diminishes the possibility of finding a difference. LeVay also reasons that viral attack is unlikely to explain his result because there were no differences in the sizes of several other areas that he measured in the hypothalamus of homosexual and heterosexual men. This is more convincing evidence, but it is still not conclusive as many viruses that attack the brain target specific regions. The encephalitis lethargica virus, for example, targets cells in a specific region called the substantia nigra. The exact state of health of each subject, as well as the cause of death, cannot therefore be ignored in any of these studies. Also, LeVay's study has not been replicated by any other researcher.

There is a similar problem with the report by Laura Allen and Roger Gorski,[27] of the University of California, Los Angeles, that the anterior commissure—a large tract of nerve cells that connects the left and right sides of the brain—is 34 percent larger in homosexual men than in heterosexual men. Again, almost all of the brains from the homosexual group had come from people who had died of AIDS, whereas very few of the heterosexual group had died from this cause. These examples illustrate that the studies give no more than an indication of a possible difference and certainly do not prove it, a fact that was lost when the information was reported by the media.

Allen and Gorski linked the claimed differences in the sizes of the brain regions to the action of sex hormones during development, the levels of these hormones being in part determined by the action of genes on the X and Y chromosomes. They believe that the hormones have a global action on many brain regions, including the higher centers to which the anterior commissure connects, and that differences in all of these cause homosexuality. (Note that it is homosexuality that is discussed as being caused, not heterosexuality.) Their result also hints at a more general constellation of cognitive differences between homosexuals and heterosexuals, a field of interest now being actively pursued by some scientists.[28]

A more recent study carried out by Dick Swaab and Albert Hofman[29] of the Netherlands Institute of Brain Research examined the size of the suprachiasmatic nucleus (another small region of the hypothalamus; fig. 3.1) in homosexual and heterosexual men. This region was found to be larger in a sample of ten homosexual men than in a sample of six heterosexual men. Swaab and Hofman controlled their study by comparing both heterosexual and homosexual postmortem samples from people who had died of AIDS. In rats and other animal species, this particular nucleus is known to control cycles of activity from day to night. Despite mentioning that experience as well as genetic and hormonal influences could have caused the size difference in this small region of the brain, Swaab and Hofman were inclined toward the notion that this region causes differences in sexual orientation and associated behaviors. They were less keen on the possibility that behaving as a homosexual could have caused the difference in size of this region of the brain. They also extended their results to an explanation of sex differences in behavior: "Functional sex differences in reproduction, gender, and sexual orientation might be based on anatomical differences in the hypothalamus."[30] They also thought that the size differences in the suprachiasmatic nucleus might cause different sleep-wake cycles in homosexual and heterosexual men.

Given the differences in lifestyle of most homosexuals and heterosexuals, there may well be differences in many of their behaviors and also in the ways in which they think, but this does not imply that these differences are caused by genes or hormones. If differences in brain size, structure, or function are conclusively found between homosexuals and heterosexuals, this would be merely a correlation that tells us nothing about what causes the differences. Researchers often ignore this fact, and falsely conclude that the claimed differences in the brain have genetic causes. For example, even though LeVay mentioned that the difference in the size of the nucleus between the homosexual and heterosexual groups may have been caused by being homosexual or heterosexual, in general he reported that he strongly believes that genetic factors are the cause of the size difference and that these, in turn, cause the difference in sexual orientation. This belief is exemplified by his statement that "the scientific evidence presently available points to a strong influence of nature, and only a modest influence of nurture."[31]

In fact, LeVay has suggested that there may be individual differences in how brains respond to sex hormones and that this may be determined by genes.[25] In this way, the hypothetical gay gene makes a brain that can either take advantage of the male sex hormones circulating in the bloodstream so that the individual becomes heterosexual, or it can deny the brain this "opportunity" so that the individual becomes homosexual because region $INAH_3$ does not grow large enough.

Social Impact of the Gay Gene

Researchers seeking to uncover a gene that codes for lesbian and male homosexual behavior justify their work by claiming that their findings may remove prejudice, as proof of a genetic cause of homosexuality would mean that the individual has no choice but to be homosexual. This would provide a strong argument against the commonly held belief that people become homosexual after being seduced by other homosexuals.[32] The effect of this might be to remove fears and restrictions on employing homosexuals in certain professions (such as in childcare and the military forces). But the discovery of a genetic cause could at best lead only to tolerance of gay men and lesbians, which falls far short of social acceptance. To give a comparison, recognition of the role of genes in determining skin color has not led to a reduction of racial prejudice. In fact, in some circles (for example, in a small group of psychologists led by J. P. Rushton,[33] of the University of Western Ontario), it has had the opposite effect, and skin color has been linked to the genetic causation of other features, including patterns of thought, all of which have been used by one group of people to oppress another. Knowing what causes a perceived difference, or having a strong belief about what causes it, does not always lessen prejudice.

The findings of Hamer and colleagues were welcomed by the gay community in some parts of the United States, because in certain states there are laws that proscribe against the victimization of individuals on the basis of genetic difference.[34] Other gay communities welcomed the news because they thought they might be accepted into Christian churches now that homosexuality was no longer seen as a matter of choice. Elsewhere, however, potentially negative uses of this information raised deep concerns.[35] Some were worried that there might be moves to alter sexual orientation away from homosexuality by manipulating an individual's genes or brain.[36] It might take a stretch of the imagination to conceive of a society in which fetuses with the Xq28 gene sequence are aborted, but some of us can still remember the practices of Nazi Germany.

Despite the social, moral, and legal speculation generated by the work of Hamer and LeVay, the scientific fact remains that these findings are inconclusive. It is therefore premature to base any policies on them at this time. To quote Aaron Greenberg, an attorney in Chicago, and Michael Bailey, of Northwestern University, "If one wishes to make moral or legal judgments based on an inference of immutability from some form of biological causation, these connections must be adequately analyzed and clarified, not merely assumed."[32]

This quotation raises the issue of the immutability of behavior. People tend to think that "biological causation" (here meaning genetic or hormonal) of a behavior makes it immutable, that is, impossible or at least difficult to change. But whether this is the case depends entirely on the behavior itself and has little, if anything, to do with the behavior having a biological cause. Some patterns of behavior learned early in life are almost impossible to change later on. The same can be true even for some behavioral patterns learned later in life, and these do not have to involve any physical reward such as those obtained by eating or drinking. Yet the notion is widespread that biological causation means that a character is immutable. This notion ignores the fact that experience and learning can alter biology, as we will see in the following chapters.

Genes on the Rampage

We have looked at examples of research in which genes are postulated to affect human behavior. As we have seen, all of these are inconclusive, even if, at the popular level, they may sound attractive. We know nothing about how such a genetic influence might work. Despite this, many people believe that genes have a strong influence on even the most complex of human behaviors and that genes are the biological basis of our societies.[37] I would argue that the pervasiveness of genetic theories of human behavior today is a reflection of conservative social values and forces, as genetic determinism implies that differences between groups are not only natural but should not and will not disappear. Genetic theories are not a reflection of new scientific facts. Despite the availability of new technologies in molecular genetics, we have no new and convincing evidence that links human behavior to the direct expression of the genes alone, as is often stated by the general media and some scientists. These voices are the echoes of a society at pains to understand itself, and to do so in the most rigid and unflinching terms. They are an effort to recruit science into the social debate and use it to uphold the status quo.

Regrettably, the search for the certainty of knowing ourselves has taken us to the most basic building blocks of our makeup. It is, of course, useful to understand these building blocks, but we must realize that, by themselves, they cannot explain the whole. When we look at the whole, we must also acknowledge its complexity. Our preoccupations with sex and with what is "normal" sexual behavior have profound effects on what research is done in this field and how the results are interpreted. When we move prejudice aside, it becomes clear that many biological descriptions of behavior are flawed. This does not mean that it is useless to apply scientific methods to understanding human behavior, but rather that we should be aware that certain subjects in the biological and psychological sciences are, as the biologist Ruth Hubbard of Harvard University has said, "suffused with cultural meanings and embedded in power relationships."[38]

4

Hormones, Sex, and Gender

In chapter 1, we looked at the debate about nature (genes) versus nurture (experience and learning) in determining gender and went on to discuss genetic influences and the effects of experience. I deliberately left sex hormones out of this debate because hormones sit at the interface of nature and nurture. On the one hand, the X and Y chromosomes determine how the gonads will develop (into either ovaries or testes) and influence which hormones they will secrete (testosterone, estrogen, or progesterone) both before and after puberty. On the other hand, the secretion of these hormones is influenced by factors from the outside environment. Certain experiences can change hormone levels.

For example, when men are under extreme stress, their testosterone levels decline and remain low for a short period of time. This change in hormone level was established by measuring the testosterone levels of U.S. soldiers in Vietnam, both when they were relatively relaxed and again just before they embarked on a mission to fight in the war. The testosterone levels were markedly lower in the latter situation.[1] It seems that stress affects the functioning of the pituitary gland and this in turn affects the amount of testosterone secreted by the testes. Sexual activity has the opposite effect: it elevates testosterone levels in men. Even thinking about sex or watching a sexy movie can have the same effect. Winning a tennis match elevates testosterone levels, whereas losing a match lowers them. This elevation of testosterone levels in the winner seems to be related to his elated mood because the elevation in testosterone depends on how the winner feels about winning a match. The rise is lower or does not happen at all in men who do not regard winning the match as important, or if winning was by chance rather than the player's own effort.[2]

These changes in the amount of testosterone circulating in the blood happen for a short time, then the levels return to normal. But other social influences can alter the secretion of sex hormones for much longer periods and even change them more or less permanently. For example, stress experienced during critical stages of development (such as puberty) can have long-lasting effects on the levels of the sex hormones circulating in the bloodstream.

Before discussing whether the sex hormones have a role in causing sex differences in the brain, we need to know something about what controls the levels of sex hormones circulating in the blood, and how these hormones are taken up by certain target tissues. The testes secrete testosterone and the ovaries secrete estrogen and progesterone into the blood. These sex hormones then circulate in the bloodstream and reach their various target tissues, which contain receptors (proteins) that recognize them. Molecules of hormone that have arrived at the cells from the blood bind to these receptors.[3] Because the hormone molecules can be bound in this way, the concentration of the hormone in the target tissue can build up to be much higher than the concentration in the blood. There are different receptors for the different hormones, each receptor matching its preferred hormone molecule like a lock and key. There are receptors for the sex hormones in the genitals and parts of the body that are physically different in women and men. There are also receptors for these hormones in the nerve cells of some parts of the brain, and these are the ones that interest us here.

Receptors in the preoptic region of the hypothalamus are essential for controlling the levels of sex hormones circulating in the blood. This part of the brain is located well below the cortex, just above the place where the pituitary gland is attached to the base of the brain. The nerve cells in this region have receptors for sex hormones, so they can respond to the levels of sex hormones circulating in the blood (fig. 4.1). Nerve cells in the preoptic region then have an influence on the anterior pituitary gland, which is located underneath the brain. So, for

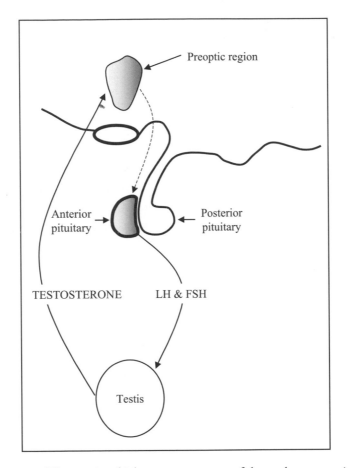

FIGURE 4.1 The way in which testosterone, one of the sex hormones, is controlled. The preoptic region of the hypothalamus (see fig. 3.1) exerts an influence on the secretion of two hormones—luteinizing hormone (LH) and follicle-stimulating hormone (FSH)—from the anterior lobe of the pituitary gland. LH is particularly important here (see text). It is released into the bloodstream and acts on the testis, causing it to secrete testosterone into the bloodstream. Testosterone then exerts a feedback influence on the preoptic region of the hypothalamus. If the concentration of testosterone in the blood is too high, the preoptic region decreases the release of LH from the pituitary gland, so the amount of testosterone secreted by the testes is lowered. The reverse occurs if the concentration of testosterone in the blood is too low. The preoptic region has receptors for testosterone, and exerts its control through neurohormones released from nerve cells into a special network of blood vessels that links the hypothalamus to the anterior part of the pituitary gland. This blood system, known as the portal system, is not shown.

example, if the level of testosterone in the blood of a male is too low, the pituitary gland is stimulated to secrete another hormone, called luteinizing hormone, into the blood. When the luteinizing hormone reaches the testes, it triggers them to secrete more testosterone. If the level of testosterone in the blood is too high, as detected by the nerve cells in the preoptic region, the amount of luteinizing hormone released by the pituitary is decreased and, in turn, the testes secrete less testosterone. This kind of control system is known as *negative feedback control*. The term "negative" is used because when one hormone is too low, the other becomes higher, and vice versa. Another hormone, called follicle-stimulating hormone, is released into the bloodstream along with luteinizing hormone. Follicle-stimulating hormone is part of the same control system and enhances the secretion of testosterone.

In women, the secretion of estrogen and progesterone by the ovaries is controlled by the same preoptic-to-pituitary loop, and the pituitary gland releases the same hormones—luteinizing hormone and follicle-stimulating hormone. But the situation is a little more complicated in females than in males because the levels of estrogen and progesterone change according to the menstrual cycle. At the beginning of the cycle, follicle-stimulating hormone released from the pituitary gland, assisted by luteinizing hormone, stimulates the ovary to produce and secrete estrogen. These pituitary hormones also stimulate the growth of follicular (egg) cells in the ovary. The estrogen then stimulates the pituitary gland to release more luteinizing hormone by a mechanism known as positive feedback, and less follicle-stimulating hormone by negative feedback. Under these conditions, the egg follicle is released from the ovary (this is ovulation), and the ovary tissue that had been surrounding the follicle (the corpus luteum) begins to produce progesterone as well as estrogen. Without fertilization of the follicle, the corpus luteum regresses and the amounts of sex hormones secreted decline until the cycle begins again.

In this way, the brain controls the amounts of sex hormones secreted into the blood. Most of the time it keeps the blood levels of sex hormones within a stable range, although these vary with time of day and, especially in women, with the time in the monthly cycle. But the brain sometimes allows the levels of sex hormones in the blood to change fairly markedly, as in the case of stress mentioned above. The brain detects events in the outside world and responds to them. One way in which it can respond is by altering the level of luteinizing hormone and follicle-stimulating hormone released by the pituitary gland. This allows the levels of sex hormones in the blood to change in response to changes in the external environment. Higher centers in the cortex of the brain process information received from the outside world and, by way of nerve cell connections to the preoptic region, these higher processes affect the control of the hormones. This is how stress or sexy thoughts modulate the levels of sex hormones in the blood. Even the way we behave can affect the levels of our sex hormones.

Sex hormones are therefore an aspect of sex differences, but they lie at the interface of nature and nurture. They can be affected by behavior and they can also affect behavior. This two-way direction of influence is important to remember because much of the thinking and writing about the biology of sex differences implies that sex hormones are biological entities that cause behavior, but not vice versa. The chain of causation is wrongly seen to be from genes to hormones to behavior, as if the hormones were merely intermediaries by which the genes exert their effects. In fact, the causal chain may work in reverse, from behavior to hormones to genes. Sex hormones cannot alter the nature of the genes, but they can determine which ones will be expressed, and so control which proteins will be made by the cells in which they act. But this is not a simple matter as there are about a hundred thousand genes and thirty thousand brain proteins, and we know very little about which ones may be affected by the hormones.

Sex Hormones and Sexual Behavior

We know from experiments using animals, rats in particular, that sex hormones circulating in the blood affect sexual behavior by acting on receptors in the preoptic region. These hormones may have a similar effect on the equivalent region in the human brain, but higher centers in the cortex of the brain also affect how responsive the nerve cells in the preoptic region will be to the hormones. These higher influences are more important in humans than in rats.

Testosterone has a more direct effect on sexual behavior in the rat than it does in primates, including humans. If an adult male rat is castrated, his sexual activity declines, even though the decline lags slightly behind the decline in testosterone levels.[4] The level of sexual activity can be restored by injecting testosterone. This means there is a direct relationship between the level of testosterone circulating in the blood and the level of sexual activity. We know that this relationship is caused by the action of testosterone in the preoptic region of the brain because sexual activity can be restored by placing a very small amount of testosterone directly into the preoptic region of the brain of a castrated rat.[5]

This direct relationship between testosterone and sexual behavior is not found in primates. If an adult male rhesus monkey that has had many sexual encounters is castrated, he shows little or no decline in sexual activity, even though his testosterone levels have declined.[6] In more complex species, therefore, sexual behavior is freed from direct control by sex hormones, and higher brain centers are more important than the preoptic region alone. Although there are few rigorous studies of sex hormones and sexual behavior in humans, as might be expected, the situation in humans appears to be the same as that in other primates and different from that in rats. The sexual libido of human males is relatively independent of the amount of testosterone circulating in the blood. Although most people believe that castration lowers sexual desire and activity, this is not the case.[7] Eunuchs may have been used to guard the harem, but many of them engaged in sexual behavior. Similarly, men castrated for medical reasons in adulthood maintain normal levels of sexual activity.[8]

In females, the situation is much the same as in males. Estrogen and progesterone are effective in stimulating sexual activity in rats, and this sexual behavior coincides with estrus, when ovulation occurs. In primates, as we know from research on rhesus monkeys and marmosets in particular, sexual interest and activity are not simply dependent on the estrus (menstrual) cycle and so are not entirely dependent on the level of hormones.[9] When a chosen male is with the female, mating may take place throughout the entire cycle, although it may be higher around the time of ovulation. Castration has relatively little effect on the level of sexual behavior in rhesus monkeys, although there may be a gradual decline in its frequency.[10] In these more complex species, the presence of an appropriate partner is paramount. Sexual behavior depends on forming partnerships rather than on the levels of sex hormones in the bloodstream. The same is true in human females: social relationships have a stronger influence than sex hormones on sexual behavior.

If any hormone modulates the level of sexual interest in women it is testosterone, a so-called male sex hormone, or other male sex hormones, which are collectively called androgens. Androgens in women come from two glands: the ovaries and the adrenal glands. In fact, as we saw in chapter 1, the secretion of testosterone from these sources in women can even exceed the secretion of testosterone from the testes in some men. As a result, some women have higher levels of testosterone than some men, so androgens are not exclusive to men and, in that sense, they have been misnamed. But what do androgens do in women? They might have something to do with sexual interest because the decline in sexual behavior that follows an operation to remove the adrenal glands can be reversed by treating the patient with testosterone, but not with estrogen.[7] Women have been treated with androgen for a wide range of medical reasons, and even low doses have been reported to increase sexual interest. Some researchers have even claimed that women are more sensitive than men to the effects of androgens.[11]

Just as women produce androgens, so men produce estrogen, from the testes and the adrenal glands. They also secrete progesterone from the adrenal glands. It is not known what effect, if any, these hormones have on the behavior of men, but we know that male brains have receptors for estrogen and progesterone. In fact, male and female brains have the same number of receptors for all the sex hormones.[12] The difference between the sexes is further blurred by the fact that, in the brain, testosterone does not act as testosterone.[13] It must be converted to estrogen once it is inside the nerve cells before it can have any effect.[14] The enzyme aromatase (also known as estrogen synthase) converts androgens into estrogens. Testosterone is sometimes transformed in brain cells, by the action of another enzyme, into another active androgen, 5α-dihydrotestosterone.

All these complexities make it very difficult to pin down any differences in behavior between the sexes. It is not a simple matter of knowing what hormones are circulating in the blood in women and men. The number of receptors for each hormone in the various parts of the brain is a critical link in the chain of events, for it is the receptors that recognize hormone molecules and remove them from the bloodstream so they can work in the brain itself. If there are only a few receptors for a particular hormone, even high levels of that hormone in the blood will have relatively little effect in the brain.

The number of receptors varies from one part of the brain to another and also from one individual to another. Two individuals might have the same levels of a hormone circulating in the blood but, if one has more brain receptors, more of the hormone will be taken up by that person's brain tissue. Also, the sensitivity of the receptors can vary. One individual may have the same number of receptors for a hormone as another, but the receptors may be less sensitive, and no amount of hormone in the blood can make up for having insensitive receptors. For example, a genetically male person with insensitive receptors for testosterone, a condition known as androgen insensitivity, does not develop facial hair or any of the other physical attributes of maleness.

Another complicating factor is that the hormone molecules in the blood can be either "free," meaning that they are separate molecules not bound to any other molecule, or bound to protein molecules in the blood plasma. Only free hormone molecules are available to bind to the receptors in the cells and will therefore be effective in the brain. There is a balance between free and bound hormone in the blood—more bound molecules become free as the brain tissue removes free molecules from the blood—but both the amounts of bound and free hormone vary from one individual to another. These complications raise the question of how we would measure the effective concentration of a sex hormone.

Measuring Hormone Levels

If we wanted to test whether there is an association between sex hormones and behavior, or the performance of certain tasks, we would need a reliable way of measuring both the behavior in question and the hormone levels. It is a fairly simple matter to take a sample of blood and to assay the level of hormone in the sample. This can be done by using radioactive substances that bind to hormone molecules in the blood. By knowing how much radioactive substance was injected and measuring the amount bound to hormone, the concentration of the hormone can be measured. These assays (called radioimmunoassays) are a little difficult to get to work accurately but, once they have been established, they can be done in large numbers. We usually measure the amount of free hormone because the radioactive substance can bind only to the free molecules.

Samples of saliva can be used instead of blood. The levels of the sex hormones, and also the stress hormones, in saliva reflect the levels in the blood fairly accurately. Taking a sample of saliva is much easier and more pleasant for the subject than taking a blood sample.

Although we can measure the amount of hormone in blood or saliva accurately and easily, there are other experimental problems. First, there is the problem of when to take the sample and what to do about fluctuations in levels that occur over very short periods of time, such as minutes or hours. Second, there is the problem that both blood and saliva are physically (in terms of body structures) quite far removed from where the hormone acts in the brain. Ideally, we would want to measure the number of receptors for the hormone in the brain, as well as assaying the blood or saliva. We would also need to know what part of the brain we were interested in because the number of receptors varies from one brain site to another. In humans, this would be very difficult to find out. So far there is no way of imaging a living human brain to see the hormone receptors, as there is for seeing changes in blood flow and nerve cell activity (described in chapter 1). The number of receptors can be established only by removing brain tissue, cutting it into sections, and using radioactive-binding techniques to assay for the receptors. It is possible to study receptors in postmortem human brains, but most experiments like this are performed using animals, rather than humans. So most of the research on the number of sex-hormone receptors in the brain has been carried out on animals, usually rats but sometimes primates and birds. The results of these studies are extended to humans, but there may be differences between the species that we do not know about.

Research into the influence of sex hormones on sex differences in human behavior, then, relies on measuring hormone levels in blood, saliva, or urine. In the past, only urine measurements were available, and these involved measuring the changed (metabolized) forms of hormones excreted in the urine. They are a further step away from the brain but have an important advantage, despite being rarely used these days. The urine sample used for assaying is taken from the total urine produced over twenty-four hours, so it represents an average measurement over the entire day, whereas blood and saliva samples give a short-term measure of the hormone at the precise time of sampling. When we want to relate the hormone level to behavioral performance, the urine measurement may be the most useful because it irons out the rapid fluctuations that can happen from minute to minute, or hour to hour, in the blood. Of course, if a particular behavior is precisely related to the short-term fluctuations in blood levels, we would miss seeing it if we used the urine sample but, so far, there is no known case of such a relationship. Most of the behavioral characteristics that scientists have attempted to link to sex hormones remain stable over longer periods of time (such as sex differences in behavior), so urinary measurements may give a better indication of what is happening. The best sample to use may depend on the particular behavior being investigated, but it is often desirable to sample the blood (or saliva) as well as the urine. Unfortunately, the fashion in research of using just the latest method has meant that few researchers today use urine samples.

I have discussed these issues only in general terms here. Hardly anything is known about the short-term compared with the long-term relationships between sex hormones and behavior, if indeed any true relationship exists at all. And we know very little about the brain regions involved in controlling any of the behaviors that have been investigated, such as spatial performance, aggression, and verbal and mathematical abilities. As we will see in the next section, links between levels of sex hormones and human behavior are more a matter of speculation than scientific fact. They are often extrapolated from the results of studies using animals. This may not be a problem if the researcher makes it clear that this is being done, but regrettably that is not always the case.

Sex Hormones and Styles of Thinking

Receptors for sex hormones are also found in parts of the brain other than the hypothalamus. If radioactive sex hormones are injected into rats, we can establish where in the brain the hormones are concentrated because the hormones bind to their receptors. In addition to regions in the lower part of the brain, including the preoptic area, the hormones are concentrated in parts of the limbic system, which has a role in the emotions, and in "higher" areas of the cerebral hemispheres of the brain, which are involved in complex processing of information. One of these areas is the hippocampus, a region of the brain involved in processing spatial information and shorter-term memory. Despite the fact that there are no sex differences in the distribution of these receptor sites, the fact that the steroid hormones concentrate in higher regions of the brain suggests the possibility that they might influence ways of thinking.

Some time ago, Donald Broverman, then of Worcester State Hospital in Massachusetts, and colleagues investigated whether the amount of testosterone circulating in the blood might affect performance of tasks in which women and men perform differently.[15] They did this by comparing men with low and high levels of androgens. The subjects were selected according to their physique, as high-androgen men have broader shoulders and chests relative to their height than low-androgen men. High-androgen men also have more hair on the body, although this varies among different ethic groups. Performance was measured on simple tests, such as speed of reading the names of colors, and speed of naming three simple figures repeated at random in a long sequence. The high-androgen men performed better on this task than the low-androgen ones. They could focus on the task without being distracted. This particular ability did not serve the same subjects as well on another task that required them to find a simple figure hidden within a much more complex one (the Witkin embedded-figure test). On this task, the low-androgen men found the simple figure sooner than the high-androgen ones.

Broverman and colleagues interpreted these results as showing that the amounts of androgens circulating in the blood determine patterns of thinking. They said that the high-androgen men expressed the "automatization cognitive style."[15] The automatized behaviors are those performed with little conscious effort because they have been well practiced and overlearned. Broverman contrasted this style of cognition to that required to perform the Witkin embedded-figure task, a "perceptual restructuring task." They said that perceptual style was specialized to perform repetitive responses accurately or to inhibit such responses to allow perceptual restructuring.

Because high- and low-androgen men have different physiques, there is another explanation for the results. Physique has a large effect on how we are raised and how other people respond to us. It is possible that the low- and high-androgen men are taught or encouraged to use different thinking styles. This means that their differing hormone levels could have affected their thinking style indirectly by changing the social environment, rather than directly by acting on receptors in the brain.

The average performance of women and men differed on the tasks used by Broverman and colleagues. Despite the differences between high- and low-androgen men, men perform better overall than women on the Witkin embedded-figure test and other similar tests, whereas women, it has been found, perform better than men on simple, repetitive tasks. At first it may seem surprising that women perform more like the high-androgen men than the low-androgen ones. Broverman and colleagues explained this by citing evidence that estrogen at high concentrations acts as if it were testosterone. From a different perspective, I would argue that it is impossible to separate the effects on thinking style of physique and the way in which individuals are treated from any potential effects of the hormones. Women and men may demonstrate different thinking styles because they are raised in different ways, again not by a direct action of the sex hormones on the brain.

We need to consider the conclusions reached about the two different thinking styles. Broverman and colleagues claim that their tasks tell us something general about thinking style, extending beyond the particular tasks themselves. The results on a few different tasks lead them to believe that they have unmasked something fundamental about how people think, and that there is specialization for either one style or the other.[16] It is not insignificant, therefore, when they conclude that women are better than men at simple, repetitive tasks, requiring automatization of cognitive style and "a minimum of conscious effort." Poor performance on the Witkin embedded-figure task has been associated with dependency in interpersonal relationships, suggestibility, conformity, and a lack of self-reliance. These characteristics are not associated with success in leadership roles. If we were to take these claims seriously, the top jobs would have to be reserved for men, and within the male group for the tall, slightly built, low-androgen men. This would be seen as their biological right.

But Broverman disagrees with this particular association between cognitive style and employment success, and has reported data showing the opposite: he found that the automatization style of thinking was more common in older men with high-status occupations and great social success.[17] He suggests that this relationship happens because these men were less easily distracted and were more aggressive than "non-automatizing" men. If the employment success of women had been included in this study, the result would not have been so clear-cut. Despite their claimed automatization style of thinking, women are not usually found in high numbers in high-status occupations. One could argue that social attitudes about men and women might override any particular advantage women might have as a result of their thinking style. But then, social attitudes might affect the success of men with different physiques, which, according to Broverman, are inextricably linked to their thinking styles because both are caused by the action of the same hormones.

In my opinion, the two separate thinking styles that Broverman and colleagues have observed might be limited to very specific tasks and have little to do with performance in a more general sense. Perceptual restructuring might, for example, be tested in many different ways, and women might perform better than men on some such tasks but not others. It is reasonable to suggest that women might be better than men in performing perceptual restructuring tasks requiring the use of language, because on average they have better language abilities than men.

Spatial Abilities

In chapter 2 we looked at the genetic theories for the claimed superior ability of men to use spatial information, about the location of places and objects. We also saw that women perform better than men on some forms of spatial tasks and that there is cultural variation in which sex performs best in the standard tests for spatial ability. Paula Caplan, of the University of Toronto, and colleagues have reviewed the literature on sex differences in spatial ability and found conflicting results between studies.[18] They concluded that there are no grounds for saying that men have better spatial abilities than women. Yet, despite these contradictions, many researchers still cling to the idea that men have better spatial abilities than women, and that this is caused by the action of sex hormones on the brain.

Doreen Kimura, of the University of Western Ontario, is one of these scientists. Together with some colleagues, she tested women and men on a task requiring them to read a map and memorize a particular route.[19] The men learned the route in fewer trials than the women but, once they had learned the route, the women remembered more of the landmarks than the men. This result was interpreted as evidence that testosterone causes superior navigational abilities, in line with genetic theories about the selection of males with better spatial abilities, used in hunting, as discussed in chapter 2. Whether the map-reading ability tested by Kimura and colleagues has anything to do with the skills used by our distant ancestors remains pure speculation, and to think that ancestral men performed these better than ancestral women is no more than imagination. The fossil and archaeological records left by ancestral humans tell us nothing about any differences in the spatial abilities of women and men.

A more direct contradiction to claims of positive association between testosterone levels and spatial abilities comes from the work of Valerie Shute of the University of California, Santa Barbara.[20]

She found that men with low levels of testosterone performed better than men with high levels of testosterone on spatial tasks. In the same study, women with higher levels of testosterone performed better than women with lower levels of testosterone. One might conclude, as Kimura has done, that there is an optimum level of testosterone, such that levels higher or lower than the optimum impair spatial performance, but this puts the claimed sex difference on shaky ground. This is because the optimum would have to lie somewhere in the range of high-testosterone women and low-testosterone men. If this were the case, it would blur any overall difference between women and men. The more we look at the results of studies on sex differences in spatial ability, the more unlikely the hormonal explanations seem, because there is no clear picture of a difference between women and men.

If the circulating levels of testosterone do affect spatial ability, then the other sex hormones, estrogen and progesterone, might also have some effect. But there is some evidence suggesting that women may perform best on spatial tests at the stage of the menstrual cycle when both these hormones are at their lowest levels.[21] The balance between all the sex hormones would need to be taken into account, but it is rare for a study to consider more than one hormone at a time.

Also, if the spatial ability of women does vary with the stage of the menstrual cycle, and this has yet to be proven, this should be taken into account in comparisons of women and men. As Walter McKeever of the University of Toledo has pointed out, if women are selected with no particular regard for the stage of the menstrual cycle, it is likely that most of the subjects will be in the phase with low levels of both hormones because this is the longest phase of the cycle.[21] This could explain why some studies have found lower spatial performance scores in women than men. Although this has not yet been proven, it raises the important point that the menstrual cycle should be taken into account in any study into the effects of sex hormones on the behavior of women.

Sex Hormones and Brain Development

As we saw in chapter 1, the ovaries or testes begin to secrete hormones very early in development, even before birth, so they can act on the developing brain. This was first shown to be the case in rats. In the early 1960s, Geoffrey Harris and Seymour Levine, then of Oxford University, injected testosterone into rat pups during the first five days after birth. The female rats were masculinized,[22] meaning that later in life they did not show the female cyclical pattern of secreting luteinizing hormone from the pituitary, as occurs in the estrus cycle (equivalent to the menstrual cycle of humans), and they showed sexual behavior more typical of males. During a sexual encounter, they showed more male-typical mounting and less female-typical crouching (the lordosis posture). The absence of testosterone during the first five days after birth had the opposite effect: male rats castrated at birth showed cyclic secretion of luteinizing hormone when treated with estrogen and progesterone or testosterone in adulthood,[23] and no amount of manipulation of the sex hormones in adult life made their hypothalamic control of the pituitary like that of a male. They also adopted the female-typical sexual behavior, crouching instead of mounting. If the castrated males were injected with testosterone during the first five days after birth, the male pattern of secretion of luteinizing hormone (that is, not a cyclical pattern) was restored, provided that sex hormones were also administered in adulthood to compensate for the castration.

These experiments show that testosterone must be present in the blood during the first few days after birth if a male-typical brain is to develop, so this is the sensitive period during which testosterone affects the differentiation of the brain into a male or female type. Strictly, we should say this applies only to the hypothalamus, the preoptic region in particular, which controls the secretion of luteinizing hormone by the pituitary gland as well as sexual behavior. These experiments have had an enormous effect on research into sex differences and the brain in other species, particularly humans. The results have been extended, perhaps incorrectly, to all parts of the brain.

So the presence of testosterone in early life switches the development of the brain from a female path to a male path. Testosterone can have this effect irrespective of whether the individual is genetically XX or XY. In the absence of this testosterone switch, the brain develops as a female type. Normally the XY genes lead to testosterone secretion, switching the brain to the male type to match the genetic type, and the

XX genes lead to the development of ovaries and the absence of testosterone leads to a female-type brain developing. Testosterone is thought to have this switching effect by acting directly on the developing brain but, as we will see in the next section, this may not be right. Testosterone may affect the brain indirectly by changing the way in which the mother, and other individuals, treat the infant.

The sensitive period, during which the testosterone has its switching effect, varies from one species to the next. For example, Robert Goy and colleagues at the University of Wisconsin showed that, in guinea pigs, the sensitive period occurs before birth.[24] This may be because guinea pigs spend longer in the womb before birth than other species of a similar size and are born at a more advanced stage of development than rats, for example. This means they can move about and are more independent of the mother than are rat pups. It is assumed that the sensitive period for the effect of the sex hormones on brain development also falls before birth in humans. But simple principles of development make this explanation unlikely. Humans are not born in an advanced stage of development. The human brain develops before birth, but much of its development continues after birth. The development is rapid in the first months after birth but actually continues for several years. Given that, on average, boys secrete higher levels of testosterone than girls both before and after birth, it seems possible that any effect of testosterone on the developing brain could happen after birth. Nevertheless, research in the 1970s indicated that exposure of female human fetuses to high levels of androgen masculinized their behavior in later life. But the levels of androgen were much higher than normal.

The exposure to androgens came about because their mothers, like many women at that time, had taken a drug called progestin to prevent miscarriage. Progestin was later found to have some androgenic action that masculinized the genitalia of some of the female children born to mothers who had taken it. John Money and Anne Ehrhardt, then both at Johns Hopkins University, took advantage of this situation to test whether exposure to progestin might have masculinized the behavior of the girls.[25] They chose progestin-exposed girls who had no obvious masculinization of the genitalia to control for any potential effects that having masculinized genitalia might have on the way the girls were treated by their family and others. The study included an additional test group of girls with adrenal hyperplasia, a condition in which the adrenal gland produces higher than normal levels of testosterone, as well as other androgens and stress hormones. These girls would have had masculinized genitalia, because that is how the condition is detected, although the researchers never mentioned this.

One procedure used by Money and Ehrhardt was to interview the girls' mothers by telephone and to ask them whether their daughters were tomboys or liked wearing girls' or boys' clothes, playing with girls' or boys' toys, and so on. Another approach was to test some of the girls directly to assess their gender identity, attitudes about work and marriage, IQ scores, and whether they had homosexual fantasies. But most of the conclusions were based on the interviews with the girls' mothers, even though it is well known that information obtained by such retrospective reporting can be unreliable, particularly when given over the telephone.

The results showed a pattern of nontraditional preference for "tomboy" play, boys' clothing, higher IQ scores, and a choice of career over marriage. But these results were compromised by inadequate controls: some of the sisters of the test subjects had the same patterns of behavior, even though they had not been exposed to progestin or high levels of androgens. A follow-up study of girls with adrenal hyperplasia by Susan Baker and Ehrhardt, both of Columbia University, was better controlled and showed that their only difference from control girls was that they played more energetically. This play was now referred to as "increased energetic play," rather than the value-loaded "tomboyishness."[26] A more recent study by Sheri Berenbaum, of the Chicago Medical School, and colleagues found that girls aged three to eight with adrenal hyperplasia preferred to play with boys' toys, whereas no such preference was expressed by their female relatives without adrenal hyperplasia.[27] But the parents of androgen-exposed girls in all these studies may have contributed to the behavioral differences as a result of their reactions to the masculinization of their daughter's genitalia or prior knowledge about the effects of the drugs.[28] A study by Frouke Slijper, of the Sophia Children's Hospital in the Netherlands, found that parents of girls with adrenal hyperplasia exaggerated their reports of their daughters' preference for energetic play, possibly because they harbored doubts about the sex of the child.[29] These doubts expressed by the parents can be transferred to the child, who responds by showing behaviors that are not so typical of girls.

Slijper also found that the cause of increased energetic play (or "romping behavior," as she called it) and other apparently masculinized behaviors was the experience of being ill in early life, rather than being exposed to abnormally high levels of androgens. Children with adrenal hyperplasia fall ill because of excessive loss of salt from their bodies, and they are also hospitalized for corrective surgery of the genitalia. Slijper compared girls and boys with adrenal hyperplasia with girls and boys who had been ill with diabetes, which does not cause abnormal levels of the sex hormones but does mean that, like hyperplasic children, sufferers faced prolonged illness and hospitalization. Compared with control girls who had no known illness, both the diabetic and the hyperplasic girls were more "boyish," whereas both diabetic and hyperplasic boys were more "girlish" than the control group of boys. So it seems that illness in early life, rather than exposure to abnormally high levels of androgens, can alter play behavior so it is no longer typical for the child's sex.

Other studies following up the suggestion that girls exposed to androgens might have lesbian tendencies found that they had more homosexual fantasies than the controls, but there was no evidence of increased homosexuality.[30] Fantasies are not reliable indicators of sexual orientation. Despite this weak evidence, Money as well as Gunther Dörner, of Humboldt University, Berlin, continued to claim that lesbian behavior could be caused by exposure of girls to high levels of androgen before birth, and that male homosexuality could be caused by insufficient exposure to androgens before birth.[31] Money went on to propose that increasing the amount of sex-hormone imbalance during development in the womb could cause homosexuality, transvestism, or transsexualism, in that order. Furthermore, because stress suffered by mothers during pregnancy can lower the level of testosterone circulating in the blood of the fetus,[32] he reasoned that stress might cause homosexuality by reducing the amount of testosterone to which the male fetus is exposed.

Dörner reported an increased incidence of homosexuality in men born during the Second World War, a result he put down to the stress suffered by their mothers during pregnancy.[33] But the recorded incidence of homosexuality is always unreliable owing to social influences on acceptability and disclosure, so reporting can vary with changing social attitudes. Attitudes about homosexuality began to become more tolerant, if not accepting, in the 1970s. People surveyed then were born during the Second World War. So an apparent increase in the incidence of homosexuality might reflect the changing attitudes of the time, rather than being an objective measure of the number of homosexuals born during the war.

Norman Geschwind and Albert Galaburda, of Harvard Medical School, believed that male homosexuality may be caused by the action of testosterone on the development of the left hemisphere.[34] Testosterone is said to suppress the development of the left hemisphere but have no effect on the right hemisphere; so they argue that a greater than normal proportion of homosexuals should be left-handed, as each hand is controlled by the opposite hemisphere. Their reasoning is that delayed development of the left hemisphere would impair the usual development of right-handedness and cause an increase in left-handedness. But there is no evidence that testosterone can act in the way they suggest or that this might have anything to do with causing either homosexuality or left-handedness.[35] Indeed, there is no consistent evidence for a greater incidence of left-handedness in homosexual men, or for unusual levels of sex hormones.[36]

Cheryl McCormick and colleagues at McMaster University in Canada interviewed thirty-two lesbians and found that most had some degree of left-hand preference.[37] This convinced her that lesbians have an "atypical pattern of hemispheric specialization" and therefore that there is "a neurobiological difference between homosexual and heterosexual women." She said this was due to lesbians being exposed before birth to an excess of male sex hormones, although there is no evidence to support this claim. There is actually no substantial evidence to support the hypothesis that lesbians have been exposed to abnormal levels of sex hormones before birth, although some researchers promote this view.[38]

Rather simplistic thinking tends to dominate much of the scientific research of sexual preference, as we saw in chapter 3. In every theory in which hormones cause lesbianism or homosexuality, there is an assumption that homosexual men are more like women than are heterosexual men, and that lesbians are more like men than are heterosexual women. This thinking is based on traditional notions about homosexuality and lesbianism that were shown long ago by psychosocial research to be incorrect. Inevitably, such simplistic thinking at the behavioral level leads to simplistic biological explanations.

Mothers, Fathers, and Gender

Human mothers react differently to girls and boys from the time of the child's birth.[39] Once the child has been assigned to either the male or female sex, she or he will be treated as either a "girl" or a "boy" both by the parents and by other people. Even mothers who are determined to treat their male and female children in the same way display a host of different responses to girls and boys. The cultural influences on the way we react to girls and boys are very persistent. This can be shown by dressing a young child at one time as a girl and at another time as a boy and recording the responses of adults to that child. When dressed as a girl, the child is encouraged to play with dolls and is spoken to frequently; when dressed as a boy, the child is encouraged to play with hammers and trucks and is spoken to infrequently.

The assignment of sex is sometimes incorrect, as is often the case for girls with adrenal hyperplasia because the enlarged clitoris is mistaken for a penis. In these cases, the child often adopts to stay with the initial designation of sex, even though both physiology and anatomy may indicate otherwise in later life. This tells us that the experience of being raised as a boy or a girl in the early period after birth has long-lasting effects on behavior.

In chapter 2 we mentioned two families that have a genetic condition preventing the conversion of testosterone to 5α-dihydrotestosterone until puberty. As a result of carrying this gene, some of the genetic males look like females until they begin to develop a penis at puberty, as the penis requires 5α-dihydrotestosterone to develop. At puberty, when the penis begins to develop, these genetic males switch to wearing boys' clothes and adopt the male gender role.[40] As they apparently switch gender identity without any difficulty, it has been suggested that they may have had "male" brains all along. Having a male-type brain, despite looking like a female before puberty, is said to be caused by the action of testosterone, which these individuals secrete normally before puberty. An alternative explanation for the ease in shifting gender might be found in the attitudes of their family members. Because the genetic condition is common in these families, the family culture may have adapted by being more flexible about gender and keeping an open mind about assigning gender until after puberty. The ease of the switch could therefore be due to either social learning or the action of hormones on the developing brain, or even a combination of both. It is impossible to tell from the research so far.

Whether gender identity is determined by early exposure to sex hormones or by learning in early life (or both) is still unknown and debated. In the 1970s, Money reported an unusual case that became widely quoted as evidence that learning is more important than the influence of genes or hormones.[41] It concerned a pair of identical male twins, one of whom suffered an unfortunate accident when a doctor was performing a minor operation. The surgical knife slipped and cut off the child's penis. The parents, acting on medical advice, decided to raise the child as a girl. The testes were later removed and estrogen therapy was given from puberty on. Money reported that the child adopted a female gender identity and behaved in ways typical of a girl, despite being genetically a boy and being exposed to androgens both before birth and until the testes were removed.

For many years this case was seen as evidence that a child is gender-neutral at birth and then learns to behave as a boy or a girl. But this claim was questioned recently. Now more than thirty years of age, the same individual was contacted again and the case reviewed. It was found that he had reverted to living as a male.[42] He had apparently felt a misfit as a girl and at fourteen years of age refused to continue taking estrogen treatment and having further surgery to construct a vagina. He was then told of his correct genetic sex, and he chose to adopt the male gender identity, receive injections of testosterone, and have a mastectomy. This outcome was seen as evidence against the neutrality of gender at birth and in favor of gender being determined by exposure to androgens before birth.

There are alternative explanations, however. Although the child may not have been told of his correct genetic sex, he was treated differently from other children because he needed hormonal treatment and surgery, which required frequent medical attention to his genital region. In addition, his parents must have been anxious, and this is likely to have transferred to the child. For these reasons, we should be cautious in using this case to try to interpret the normal development of gender role and identity. It is an abnormal situation that may shed no light on gender development.

As far as humans are concerned, there is no conclusive evidence that hormones play a greater part than learning in determining gender identity. We might therefore consider turning to experiments using animals to help us out. There is evidence that mothers treat female and male offspring differently, at least in rats, ferrets, and primates. Celia Moore, of the University of Massachusetts, and colleagues have shown that mother rats lick their male pups in the region of their genitals and anus (the anogenital region) more than their female pups,[43] and the same is true in gerbils and ferrets.[44] This preferential licking of males affects their sex-typical behavior as adults, and it also affects their physiology. Moore gave the female pups extra anogenital stimulation by stroking their anogenital region with a paintbrush several times a day. The females behaved like males as adults: they mounted other females, showed patterns of activity more typical of males, and even secreted luteinizing hormone from the pituitary in the continuous pattern typical of males, instead of the cyclic pattern typical of females.[45]

The male pups receive more licking from their mothers because they have an attractive substance in their urine.[46] The higher levels of testosterone cause the preputial gland to secrete this substance into the urine. The mother can smell this, and this attracts her to lick the males more than the females. If the mother is prevented from smelling the urine, she no longer distinguishes between her male and female pups.[47] Female pups treated with testosterone receive more than normal amounts of anogenital licking and behave like males when adult, whereas castrated male pups receive less licking than normal and behave like females as adults, and even develop the female pattern of cyclic secretion of luteinizing hormone from the pituitary gland. Normal males that receive extra stimulation of the anogenital region also develop larger testes than those receiving a normal amount. The anogenital licking also stimulates growth of a region of the brain, which develops more nerve cells.[48]

These remarkable results show that sex differences in physiology, behavior, and even brain structure depend on maternal stimulation of the pup's anogenital region in early life. The same may be true for other species, with maternal stimulation affecting development and behavior in later life. Different species may have different forms of stimulation, but anogenital licking is important in several different species. In natural settings outside the laboratory, both parents may contribute to gender development by providing different amounts of attention to female and male offspring.

The fact that mother rats pay different amounts of attention to the anogenital region of their female and male pups makes us reconsider the earlier experiments of Harris, Goy, and Phoenix (whose research has been mentioned in the previous section). Certainly, the presence or absence of testosterone in the early period after birth determines sex differences in behavior. As we have already seen, the presence of testosterone leads to male-typical behavior and its absence leads to female-typical behavior. There is no doubt that the observations made by Harris, Goy, and Phoenix are correct, but Moore and colleagues have demonstrated that the route by which the testosterone affects behavior may be via the mother. The mother rat responds to the presence of testosterone by being attracted to the pups' urine, and it is the change in her behavior, not the direct action of the testosterone on the pups' brains, that may affect the development of sex differences in the pups.

Testosterone therefore has a role in the sequence of events, but it may be indirect. This is a radically different interpretation of earlier results and has important implications for the way we think about the development of sex differences in humans.[49] Yet little attention has been paid to the findings of Moore and colleagues.[50] It is as if people are unable to change their minds to accept the new concept.

Biology is involved even in the indirect route of altering the mother's licking behavior because the mother is responding to chemical signals in the urine. But the route of causation is not a simple linear one from the pup's genes to its hormones to its brain and then to its behavior. The development of sex differences in the rats' behavior is not orchestrated by an innate program but involves the interaction of developmentally specified events (determining the odor of the urine) and modification by experience (anogenital licking).

Much of the thinking about the role of hormones on the development of the human brain has been based on the original research with rats and the interpretations of the results made at that time. This thinking has remained unchanged despite the new evidence for the role of the mother in affecting the changes in behavior. It seems, then, that there is a reluctance to reject hormonal theories for the sex differences, and that there is a vested interest in the idea that hormones control our lives. A review of the scientific literature on testosterone and aggression led John Archer, of the University of Central Lancashire (UK), to doubt whether there is any direct relationship between the level of testosterone in the bloodstream and aggressive behavior,[51] yet the idea that testosterone causes aggression persists among scientists and the public.

Experience, Interactions, and Change

In previous chapters, we have looked at explanations for sex differences in behavior that have given either genes or hormones a primary and overriding role. These are called reductionist accounts because they simplify (reduce) complex interactions to a single causal event. Developing organisms respond to a diverse range of influences from the environment, but these are either partially or completely ignored by reductionism. At the most basic level, there are influences from the chemical environment in which the organism develops from its earliest stages. Sensory stimulation and other inputs from the surroundings then influence the organism's development both before and after birth. At a higher level, there are effects of the psychosocial environment, which changes continuously throughout life. All these influences from the environment are not separate from the organism's biology; in fact they can radically change it. Inputs from the outside determine which genes will be expressed and modulate the levels of hormones secreted into the bloodstream. The intricate processes involved in the development of a living organism cannot be separated one from another. Genes, hormones, and inputs from the outside all interact. They are completely intertwined.

The genes are not the code of life and the hormonal mix is not a "blueprint" for development, as some have claimed (see chapters 3 and 4). The effects of genes and hormones should not be seen in isolation from external inputs to the developing system. Similarly, inputs from the environment (experience) should not be considered in isolation from genetic or hormonal factors. No single factor has primacy in directing development. A genuine understanding of the processes of sexual and gender development will come about only by looking at the interactions among all these factors. This principle applies to all aspects of sex or gender difference, from the molecular or cellular aspects of the brain itself to the manifestation of brain function as behavior. It is also important to look not just at differences that might take place at one period in time (at one stage of a person's life) but also at the development of these differences.

The nature of research that can be conducted on development in humans is limited for important ethical reasons. Much of our understanding of the interactions between biology and experience has come therefore from research using animals, and we will consider two examples of this to illustrate the approach. The first comes from the laboratory of Victor Denenberg at the University of Connecticut, where experiments are being conducted on rats to examine the interaction of sex hormones and experience in the development of an important structure in the brain, the corpus callosum. The second comes from my own laboratory and examines the interactive roles of sex hormones and experience in the development of some sex differences in the chicken.

Hormones, Experience, and the Corpus Callosum

In the human brain, the corpus callosum is a large tract of nerve cells, or neurons, connecting the hemispheres. In a postmortem human brain it can be seen as a large white mass between the hemispheres and is revealed by pulling each hemisphere sideways a little from the midline, or slicing it as shown in figure 5.1. It allows one hemisphere to communicate with the other, either to share information or to allow one hemisphere to suppress the activity of the other. This suppression makes one hemisphere dominant over the other,[1] if only for a short time. The corpus callosum is believed to have an essential function in lateralization of brain function, by which we mean that one hemisphere carries out certain kinds of processing and controls certain

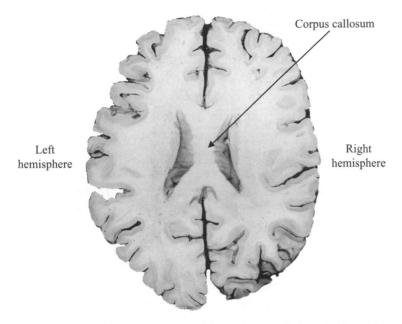

FIGURE 5.1 The human brain viewed from above and about half-way down. This shows the large tract of connections between the two hemispheres, called the corpus callosum. The front of the brain is at the top of the page, and the back of the brain is at the bottom. Note the asymmetry in the structure of the left and right hemispheres. The left hemisphere projects backwards more than the right hemisphere, and the front of the right hemisphere is larger than the same region of the left hemisphere.

functions, while the other hemisphere carries out other kinds of processing and controls other functions.

Lateralization of the brain is of interest here because there have been many suggestions that the degree of lateralization differs between men and women. Most have claimed that men have more strongly lateralized brains than women, but there have also been suggestions of the converse.[2] Whether female brains are considered to be more or less lateralized than male brains seems to depend on the function or behavior being investigated. It is perhaps better to say that the brains of women and men are differently lateralized, but we should recognize that there is a great deal of similarity rather than difference. Almost two decades ago, Marcel Kinsbourne, of the Hospital for Sick Children, Toronto, and a leading researcher in the field of lateralization, realized that differences between women and men tend to be exaggerated: "Under pressure from the gathering momentum of feminism, and perhaps as backlash to it, many investigators seem determined to discover that men and women 'really' are different. It seems that if sex differences (e.g., in lateralisation) do not exist, then they have to be invented."[3]

More recent reviews of research into lateralization in men and women have also come to the conclusion that there is little consistent evidence for sex differences in brain lateralization in humans.[4] Nevertheless, brain lateralization continues to hold a central position in debates about sex differences.

The corpus callosum mediates differences in function between the hemispheres and so generates lateralization. It is not surprising, then, that researchers have measured and compared the size of the corpus callosum in female and male brains. The results of such comparisons have been just as controversial as those examining sex differences in lateralization.

The corpus callosum is measured in terms of the area of its cross-section at the midline of the brain. This cross-section can be obtained by either slicing in half a postmortem brain preserved with formalin or by using the technique of functional magnetic resonance imaging (fMRI) described in chapter 1. Both methods have inaccuracies. Preserved postmortem specimens shrink, which may distort the measurements or generate differences that would not be present in living tissue. The magnetic resonance imaging technique has uncertainties because the computer samples the brain many times before it constructs the image to be displayed on the screen. But it has the advantage of being made on living tissue so there is no shrinkage. It would actually be best to compare results obtained from living and preserved tissue, but this is not really possible.

Interest in the size of the corpus callosum in relation to brain lateralization was renewed in the early 1980s, particularly with respect to handedness and sex differences. The interest in handedness stems from the fact that each hand is controlled by the opposite hemisphere and preferred use of one hand reflects brain lateralization. The studies conducted in the 1980s used preserved brains and subdivided the corpus callosum into various regions that were measured separately, but some errors were introduced because there are no definite landmarks in the corpus callosum for making the subdivisions.

C. de Lacoste-Utamsing and Ralph Holloway, of Columbia University, reported that the area of the splenium (the back part of the corpus callosum) was larger in women than men, a result that was later also found by Laura Allen of the University of California, Los Angeles, and colleagues.[5] The sex difference in size was considered to be due to women having a lesser degree of lateralization for processing spatial information than men. Their reasoning was that women might have more connections between those parts of the hemispheres that are interconnected through the splenium of the corpus callosum and so may be less lateralized. But they provided no evidence to support this speculation, and none has emerged since. Nevertheless, many writers in the scientific and popular media have persisted with this idea and extended it to other behaviors with no proven connection to brain lateralization or the corpus callosum. For example, an article in a popular magazine published in 1995 stated that a larger corpus callosum in women might explain why they are more intuitive than men.[6]

Sandra Witelson, of McMaster University in Canada, also claims to have found a sex difference in preserved brains.[7] She reports that the sizes of all the regions of the corpus callosum apart from the splenium are larger in men who do not have a strong hand preference than in men who are consistently right-handed. This handedness effect was not found in the brains of the women in her study. The validity of these results is in some doubt, however, because other studies using postmortem brains have failed to replicate them,[8] and studies using fMRI have failed to find any significant effects of handedness or sex on the size of regions of the corpus callosum.[9] This may be because the size difference is not present in living tissue, or because the hand preferences of the subjects were assessed in different ways in the different studies. But any indications of a sex difference in the size of the corpus callosum have not been substantiated.

Katherine Bishop and Douglas Wahlsten of the University of

Alberta in Canada recently reassessed all the data collected before 1997 and concluded that there is no meaningful sex difference in the size of the corpus callosum.[10] The whole corpus callosum is larger in men than in women but, on average, men also have larger brain size and body size than women. The sex difference disappears when these two factors are taken into account. Laura Allen and colleagues reported that the splenium region of the corpus callosum was more bulbous in females than in males, but the review by Bishop and Wahlsten found no evidence to support this. They went on to say that a radically new technique might, in the future, find a small sex difference in the corpus callosum, but qualified this by stating: "Whether the quest for such minuscule sex differences is worth the cost of time on an MRI machine is debatable, but additional studies of sex differences using samples one-tenth the appropriate size will contribute correspondingly little to our understanding of brain function."[10]

But even if a minuscule sex difference in the size of part of the corpus callosum were definitely found, it might have no significant effect on brain function. At the very least, it would be premature to claim that there was an association between the size differences and the complexities of sex differences in behavior. As Holly Fitch and Victor Denenberg, of the University of Connecticut, have said, there is confusion because many researchers have not analyzed their data correctly and have failed to account for age and hand preferences of their subjects.[11]

Research using animals might shed some light on the controversies surrounding sex differences in lateralization and the size of the corpus callosum. Here the work of Victor Denenberg and colleagues at the University of Connecticut on the development of the corpus callosum in rats is important for two reasons. First, by using an animal model it has been possible to carry out a series of controlled experiments that have revealed events important in development. Second, like humans and many other species, rats have lateralization of the brain and so can serve as a model to uncover basic principles likely to apply to the human brain. Rats have a corpus callosum that connects the two hemispheres together, although its size relative to the hemispheres is much smaller than in humans. The corpus callosum in the rat appears to be involved in lateralization, as it is in humans.[12]

The cross-sectional area of the rat corpus callosum can be determined in postmortem brains, as it also can in humans. Denenberg and colleagues found that the corpus callosum is larger in male than in female rats.[13] The development of this sex difference has something to do with the levels of sex hormones circulating in young rat pups and with experience in the first few weeks after birth.[14] The size of the corpus callosum in females increases if they are castrated just after birth (their ovaries are removed, which lowers the amount of estrogen in the bloodstream). The corpus callosum can be made larger in males by regular "handling" in their early life, which involves taking each pup away from its mother for about three minutes each day, placing it alone in a small container for this time, and then returning it to its mother. If the pups are handled each day for the first twenty days of their lives, the corpus callosum enlarges and brain lateralization is also affected. When they are adults, the handled males show a greater degree of lateralization than those that have not been handled. This handling has no effect in females unless they are also treated with testosterone. Males, of course, secrete testosterone themselves, which explains why the experience of handling alone is sufficient to increase the size of the corpus callosum of the male. As would be expected, the corpus callosum is small in castrated males.

We conclude from these results that the experience of handling must interact with the presence of testosterone to increase the size of the corpus callosum, although we do not know what aspect of handling is responsible. It could be the effect of the daily brief isolation experienced by the pups or a change in the mother's response to the pups when they are returned to her. William Smotherman, of the State University of New York, Binghamton, has shown that a mother licks her pups more after they have been returned to her.[15] As we saw in chapter 4, licking of the anogenital region by the mother is known to bring about physiological and behavioral changes that distinguish males from females. Male pups receive more anogenital licking than female pups, and artificial stimulation of the anogenital region of female pups causes them to develop physiological and behavioral characteristics more typical of males than females. So it is possible that handling has its effect on the developing corpus callosum by changing the pattern or amount of licking that the pups receive from their mother. Rat pups raised normally by their mothers, without handling or hormone treatment, could perhaps develop the sex difference in the size of the corpus callosum because the mothers lick the males more than the females. It is also possible that there are cumulative effects of testosterone and anogenital licking by the mother on the developing corpus callosum, but this has not yet been proven.

Fitch and Denenberg do not agree with me on this idea because the sex difference in the size of the corpus callosum can be seen in pups as young as three days old.[11] I believe that anogenital licking could have an effect on the development of the corpus callosum as early as three days after birth because nerve connections can grow rapidly at this stage and other changes (such as myelination of the nerve axons) can also be rapid. Whether this is correct is not known, but this hypothesis could be tested. Of course, testosterone is also likely to have a direct effect on the developing brain tissue. Licking may be associated with other forms of attention given by mothers to male pups more than female ones, but this too remains to be investigated. In a natural context, the father may also contribute to the different experiences of male and female pups.

Regardless of exactly how handling affects the pups, these results demonstrate a very important principle of development. They show that the development of the brain (in this case, the corpus callosum) depends on the combined effects of experience in early life and the particular hormonal condition of the pup at that time. Sex differences emerge because males and females have different hormones in the bloodstream during the early period after birth, but the effects of experience must interact with this hormonal influence.

There is no reason why the same principles of development should not apply to the human brain, although the exact mix of hormones and the particular form of stimulation may be different from that effective in rats. So we would expect to find that some aspects of brain development in humans depend on an interaction between experience and the levels of some of the sex hormones in early life. There might even be differences in brain structure or the way parts of the brain are connected to each other as a result of this interaction. There would be much more variation between individual humans than between laboratory rats, as the rats are bred to be similar to each other and are exposed to fairly standard rearing conditions. These variations might override any tendency for sex differences, and might even explain why data collected for the size of the corpus callosum vary from one study to the next. The presence or absence of a difference in size might depend on the subjects selected for study and their experiences early in life. If this were the case, it would be inappropriate to conclude from any study that there is an immutable sex difference.

The age of the subjects should also be taken into account. Although the interaction between hormonal condition and experience in early life is most important in determining the size of the corpus callosum, these same factors might also have some effect in later life. Even in adults the size can vary. Researchers in Denenberg's laboratory have found that the sex difference in the size of the corpus callosum of humans varies with age.[16] These age-dependent changes might also result from interactions between hormonal condition and experience. The corpus callosum, and possibly many other regions of the brain, would be responsive to the effects of hormones and experience throughout an individual's life; we say that the brain remains "plastic." Just because there are differences in brain structure (size and connectivity), it does not mean that they are caused by genes or hormones alone, or that they will not change with age and experience.

Hormones, Experience, and Visual Connections

Stimulation from the mother, and perhaps also the father, has an important influence on the development of sex differences in behavior. So what would happen if individuals were raised without their parents or other adult members of their own species? Would sex differences develop and, if they do, would this mean that they are caused by genes or hormones without any influence by the environment? My own research has addressed these questions in a general way, using the chicken as a model species because chicks can be hatched in an incubator and raised without contact with their parents or other adults. Under these circumstances, chicks do develop sex differences in behavior, which appear even in the first weeks of life.[17] For example, females adopt a pattern of searching for food that allows them to switch attention from one type of grain to another very readily, whereas males tend to focus their search on one type of grain. Chicks also have lateralization of the brain, with the left hemisphere specialized for learning to recognize food and the right hemisphere specialized for controlling sexual and aggressive behavior. I found that the degree of some forms of lateralization is less in females than males.[18] Although these sex differences develop in chicks hatched in an incubator and raised without parents, they are not determined solely by the genes or the hormones. Experience has an important part to play.

Detailed investigation of the factors involved in the development of brain lateralization in the chick has revealed interactions between genes, sex hormones, and experience. The key to the puzzle lies in the fact that, just before hatching, almost all chick embryos are oriented inside the egg with a leftward twist of the body (fig. 5.2). This means that only the right eye is exposed to light entering the shell because the left eye is occluded by the embryo's own body. This happens at the stage of development when stimulation by light can first trigger responses in the hemispheres of the brain, starting about three days before hatching. Because the left eye is occluded by the embryo's own body, this asymmetry of exposure of the eyes to light causes some forms of lateralization to develop in the chick.

The essential role of light stimulation in the development of lateralization for learning to recognize food is shown by incubating eggs in the dark. Chicks that are not exposed to light until after hatching are not lateralized, and this holds for both males and females.[19] Light promotes the development of more connections of the right (exposed) eye to the hemispheres,[20] so structural asymmetry develops in the connections involved in processing visual information in the brain. Only a couple of hours of exposure to light is needed for this to come about.

Although both sexes of chicks have asymmetry of the visual connections, this asymmetry is greater in males than in females, even when both receive exactly the same amount of exposure to light.[21] This is the effect of the sex hormones. Before hatching, male and female embryos have different levels of testosterone, estrogen, and progesterone circu-

Right eye
stimulated by light

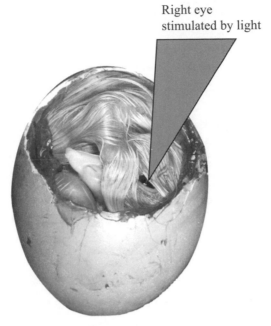

FIGURE 5.2 Just before hatching, the chick embryo turns within the egg such that its left eye is covered by the embryo's body but the right eye is next to the membranes and shell of the egg. Light can pass through the shell and the membranes to stimulate the right eye only. As explained in the text, this causes an asymmetry to develop in the visual pathways, but the sex hormones are also important in the process.

lating in the bloodstream. These hormones interact with the light just before hatching, generating the sex difference. A high level of either estrogen or testosterone promotes growth of the projections on both sides of the brain, and no asymmetry results.[22] In normal development, male embryos have high levels of testosterone before the important last few days before hatching, when they decrease. This period of low testosterone levels allows the light to generate the asymmetry. The levels of testosterone then rise again. In females, the levels of testosterone are lower than in males, but estrogen levels go on steadily increasing throughout the embryo's development and are high before hatching. These high levels in female embryos render their visual connections less responsive to light stimulation, so they develop a lesser degree of lateralization than males.

In the chick, then, the sex hormones interact with the experience of light stimulation. There are certain similarities to the interaction between testosterone and handling in the development of the rat corpus callosum but, unlike the additive effects of handling and testosterone on the size of the corpus callosum, light stimulation generates asymmetry in the chick only when the levels of steroid hormones are low enough. Comparisons such as these help us to understand the basic mechanisms involved in sexual differentiation and show that the development of sex differences in animals is a much more complicated process than we used to think. Reductionist explanations in terms of genes alone, hormones alone, or experience alone do not suffice for animals or humans.

Implications for Sex Differences in Humans

These experiments using rats and chicks demonstrate the interaction between hormonal condition and experience. Controlled research using animals has made it possible to unravel some of the complex ways in which genes, hormones, and experience influence the development of the brain and behavior. In fact, these forms of influence are not entirely distinct. Experience can alter the biology of the brain, in terms of its structure and cellular functioning, as well as the actual secretion of hormones, and biological factors can influence how experience is channeled and processed. The interaction between developmentally specified and developmentally modifiable events occurs in complex ways at every stage of development.

It is clear that genes have a role in determining the secretion of sex hormones, but this role is not absolute and, in the example of the developing chick, genes are most likely to determine that the embryo orients itself with a leftward twist inside the egg. There are, of course, many other ways in which genes exert their influences, but they always interact with other biological factors and experience. At no stage in the process of development is an event entirely caused by genes, entirely caused by sex hormones, or entirely the result of experience. Nor is it possible to see the effects of genes and experience as additive in any simple way. The rat and chick models show that these various contributions to development are inseparable at all stages.

Development of Differences

Understanding sex differences in behavior and in the brain depends on knowing about the processes behind their development. Behavioral development is complicated because many factors interact and dynamical systems are involved, changing continuously at every step in the process.[23] This is as true for animals as it is for humans. Some of the factors are thought of as being internal (genes and hormones) and others as external (inputs from the environment) to the individual, but there is no clear separation of internal and external factors because they interact.[24] For example, genes can affect the production of hormones, which can alter physical appearance, which affects how other people react to us, and this in turn affects our hormones and also our genes (as it affects which genes are expressed). This is a loop of stepwise causations, but even this is a simplified way of looking at the interactions that occur. Other factors from the environment influence each step of the process, and different interactions occur at different stages of development.

For a complete understanding of any aspect of development, we would need to know about all the factors that have contributed to the dynamical system. At the very least, any study of the development of individual and group differences (such as sex differences) would be much more valuable if it looked at all the main factors that may contribute to the development of the differences, rather than focusing on just one factor. In previous chapters, we have looked at studies of sex differences that focused only on genes, only on hormones, or only on experience. These are severely limited when it comes to understanding the development of sex differences. It is all too common for biologists and psychologists to focus on genes or sex hormones (often only a single hormone) and to make only passing reference to the potential effects of experience. As a result, little progress can be made toward understanding the interactions that occur. A lopsided view is inevitable.

This could be avoided if broader and more detailed investigations were conducted taking into account all, or most, of the potential factors that might influence sex differences.

Some researchers in the field of human gender development are beginning to recognize the complex interactive processes involved in the differentiation of gender and sex differences in behavior,[25] but others still adhere to the primacy of genes or hormones. Others see genetic and hormonal influences on the development of gender as primary in early development but accept that environmental influences have an increasing role after birth.

June Reinisch and colleagues, at the Kinsey Institute for Research in Sex, Gender, and Reproduction at Indiana University, see the interaction of biological and environmental factors in terms of a "multiplier effect."[26] In their view, genetic and hormonal differences cause sexual differentiation of the genitalia and the brain before birth and lead to relatively small behavioral differences between the sexes just after birth. As they see it, from birth onwards, behavioral differences between the sexes are "augmented by successive interactions between the individual and the social environment." They say that at puberty, when the physical differentiation of the sexes increases, social expectations and social interactions magnify the behavioral differences between the sexes, "further enhancing sex differences in behavior." This is an excellent framework in which to conduct future research, although we will understand these interactive factors only if they are assessed together in the same studies.

The brains of women and men may function differently in some ways and the same in other ways. Knowing what the differences are does not, in itself, tell us anything about what causes them. As I have tried to explain, there are several reasons for sex differences in the molecular and structural aspects of the brain, as well as in behavior, and these can be described at many levels. We cannot see the complete picture by working at only one level. Sex differences in behavior, for example, cannot be reduced to explanations purely in terms of molecular genetics or patterns of nerve cell activity inside the brain.

The methods of science tend to intercept living processes and freeze them at a moment of time, or over a brief sequence of time. For example, nerve cell activity is frozen at a moment of time as an image on a computer screen. We are inclined to focus on an isolated segment of a process of life that has both a history and a future,[27] and we attempt to construct a complete picture of the development from snapshots taken at various times through development. But it is difficult to trace the history of events that might have caused a difference that is seen at one moment of time. This is probably why so many researchers simply look at the difference and then speculate about its causes, rather than investigate them. And it is here that they fall back on ideology.

Speculations about causes are shaped by the practices and views of society, and should not be seen as objective facts simply because they are stated by scientists. Scientists have always both reflected and reinforced the attitudes of society. In the past, they have had an active role in "sexing the brain," and many of them continue to do so today. Others are attempting to take into account all of the interacting influences on the development of differences between women and men, and by doing so they are enlightening social attitudes rather than perpetuating them.

In this book I have been critical of simplistic genetic and hormonal explanations for sex differences in brain structure and function. I have drawn attention to the effects of experience, but not in isolation from genetic and hormonal influences, and have recognized the dynamic aspects of biology and behavior. Living processes are never static, and this applies to the biological processes associated with sex differences. No matter what kind of sex difference has been measured, the difference can exist only at one point of time. The individuals who have been tested, and so have provided the information about any particular sex difference, are constantly changing because they are in constant interaction with their environments. We actively select and change our environments and, at the same time, we are actively selected and changed by them.[27] Flexibility characterizes all levels of biology and behavior. In other words, our biology does not bind us to remain the same, as implied by simplistic genetic and hormonal interpretations of our behavior. We have the ability to change, and the future of sex differences belongs to us.

Recommended Reading

Becker, J. B., S. M. Breedlove, and D. Crews, eds. *Behavioral Endocrinology*. Cambridge: MIT Press, 1992.

Bleier, R. *Science and Gender*. New York: Pergamon, 1984.

Colapinto, J. *As Nature Made Him: The Boy Who Was Raised as a Girl*. New York: HarperCollins, 2000.

Connell, R. W. *Gender and Power: Society, the Person, and Sexual Politics*. Palo Alto, Calif.: Stanford University Press, 1987.

Fausto-Sterling, A. *Myths of Gender*. New York: Basic Books, 1985.

Hoyenga, K. B. and K. T. Hoyenga. *Gender-Related Differences: Origins and Outcomes*. Boston: Allyn and Bacon, 1993.

Hubbard, R. *Profitable Promises: Essays on Women, Science, and Health*. Monroe, Me.: Common Courage, 1995.

Keller, E. F. *Reflections on Gender and Science*. New Haven: Yale University Press, 1985.

Reinisch, J. M., L. A. Rosenblum, and S. A. Sanders. *Masculinity/Femininity: Basic Perspectives*. New York: Oxford University Press, 1987.

Rose, S. *Lifelines: Biology, Freedom, Determinism*. New York: Oxford University Press, 2000.

Sayers, J. *Biological Politics: Feminist and Anti-Feminist Perspectives*. London: Tavistock, 1982.

Tavris, C. *The Mismeasure of Woman*. New York: Simon and Schuster, 1992.

Tobach, E. and B. Rosoff. *Challenging Racism and Sexism*. Genes and Gender 7. New York: Femminist/City University of New York, 1994.

Vines, G. *Raging Hormones: Do They Rule Our Lives?* Berkeley: University of California Press, 1994.

van den Wijngaard, M. *Reinventing the Sexes*. Bloomington: Indiana University Press, 1997.

Notes

1. New Methods, Old Ideas

1. There are several ways to explain the interaction between genes and environment. See S. Oyama, *The Ontogeny of Information: Developmental Systems and Evolution* (rev. ed., New York: Cambridge University Press, 2000); S. Rose, *Lifelines: Biology, Freedom, Determinism* (New York: Oxford University Press, 2000).

2. Translated from G. LeBon, "Recherches anatomiques et mathématiques sur les lois des variations du volume du cereveau et sur leurs avec l'intelligence," *Revue d'Anthropologie*, 2d ser., 2 (1879): 27–104 (quotation from 60–61). Also cited by S. J. Gould, *The MisMeasure of Man* (New York: Norton, 1981), 104–105.

3. Discussed further by Gould in *The MisMeasure of Man*.

4. C. Darwin, *The Descent of Man* (London: Murray, 1871), 569; rpt., Great Minds series (New York: Prometheus, 1997).

5. F. P. Mall, "On Several Anatomical Characteristics of the Human Head Said to Vary According to Race and Sex, with Especial Reference to the Weight of the Frontal Lobe," *American Journal of Anatomy* 9 (1909): 1–32.

6. For more details see J. Sayers, *Biological Politics: Feminist and Anti-Feminist Perspectives* (London: Tavistock, 1982), 90–93.

7. This idea was used in C. Hutt, *Males and Females* (Harmondsworth, Eng.: Penguin, 1972).

8. K. Dalton, *Once a Month* (6th ed., Sussex, Eng.: Hunter House, 2000).

9. For more examples of biological explanations used to keep women out of the workforce, see G. T. Kaplan and L. J. Rogers, "The Definition of Male and Female: Biological Reductionism and the Sanctions of Normality," in S. Gunew, ed., *Feminist Knowledge: Critique and Construct* (New York: Routledge, 1992).

10. D. Kimura, "Sex Differences in the Brain," *Scientific American* 267 (September 1992): 81–87 (quotation from 87).

11. This technique is referred to as functional magnetic resonance imaging (fMRI) because it allows the working, functioning brain to be visualized.

12. R. J. Haier and C. P. Benbow, "Sex Differences and Lateralization in Temporal Lobe Glucose Metabolism During Mathematical Reasoning," *Developmental Neuropsychology* 11 (1995): 405–414.

13. R. C. Gur et al., "Sex Differences in Regional Cerebral Glucose Metabolism During Resting State," *Science* 267 (1995): 528–31.

14. B. A. Shaywitz et al., "Sex Differences in the Functional Organization of the Brain for Language," *Nature* 373 (1995): 607–609; K. R. Pugh et al., "Cerebral Organization of Component Processes in Reading," *Brain* 119 (1996): 1221–38.

15. The hemispheric differences are an example of brain lateralization. There are many other reported sex differences in the degree of brain lateralization. See also J. L. Bradshaw and L. J. Rogers, *The Evolution of Lateral Asymmetries, Language, Tool Use, and Intellect* (San Diego: Academic Press, 1993); J. B. Hellige, *Hemispheric Asymmetry: What's Right and What's Left* (Cambridge: Harvard University Press, 1993).

16. D. Kimura, "Sex Differences in the Brain," 81–87. The original report of this work was Kimura, "Sex Differences in Cerebral Organization for Speech and Praxic Functions," *Canadian Journal of Psychology* 27 (1983): 19–25.

17. Electroencephalographic recording (EEG) involves placing electrodes on the scalp in various places and using them to measure electrical activity in the brain.

18. A. Kertesz and T. Benke, "Sex Equality in Hemispheric Language Organization," *Brain and Language* 37 (1989): 401–408.

19. Spatial ability is usually tested by pen-and-paper tests requiring matching of rotated figures, as used in many IQ tests. It is usually assumed that performance on such tests translates into spatial abilities used in the "real" world, in everyday living, but these are rarely tested. In fact, as discussed in chapter 2, women perform better then men on at least one type of spatial task.

20. R. W. Connell, *Gender and Power: Society, the Person, and Sexual Politics* (Palo Alto, Calif.: Stanford University Press, 1987); S. Riger, "Epistomological Debates, Feminist Voices: Science, Social Values, and the Study of Women," *American Psychologist* 47 (1992): 730–40.

21. This topic is expanded on in M. van den Wijngaard, *Reinventing the Sexes* (Bloomington: Indiana University Press, 1997).

22. S. Oyama, "Constraints on Development," *Netherlands Journal of Zoology* 43 (1993): 6–16.

2. What Causes Sex Differences?

1. Chromosomes are strings of genes and can be seen under the microscope in the nucleus of a cell just before it starts to divide into two. Each chromosome is paired with another. The X and Y chromosomes are called the sex chromosomes because they differ between the sexes; the rest of the chromosomes are called autosomes. Females have two X chromosomes, whereas males have one X and one Y

chromosome, so the genetic sex of an individual can be seen by simply examining the chromosomes under the microscope. The sex chromosomes were discovered in 1956: see J. H. Tijo and A. Levan, "The Chromosome Number of Man," *Hereditas* 42 (1956): 1–6. Recent research has isolated a small portion of the Y chromosome (called *SRY*) that determines what sex genitalia will be: Y. W. A. Jeske, J. Bowles, A. Greenfield, and P. Koopman, "Expression of a Linear *SRY* Transcript in the Mouse Genital Ridge," *Nature Genetics* 10 (1995): 480–82.

2. The role of sex hormones (or gonadal hormones as they are sometimes called) in the development of the genitalia was discovered in the 1950s: A. Jost, "Problems of Fetal Endocrinology: The Gonadal and Hypophyseal Hormones," *Recent Progress in Hormone Research* 8 (1953): 379–418. The XY (male) genotype causes the developing gonads to secrete testosterone and Muellerian-inhibiting factor. The testosterone stimulates the growth of the Wolffian ducts, which develop into the male reproductive tracts, and the Muellerian-inhibiting factor causes the regression of the Muellerian ducts, which would have been the basis of the female reproductive tract. The XX (female) genotype does not produce either of these hormones, so the Muellerian ducts develop into the female reproductive tracts while the Wolffian ducts degenerate.

3. A. Greenfield and P. Koopman, "*SRY* and Mammalian Sex Determination," *Current Topics in Developmental Biology* 34 (1996): 1–23. Also see P. Koopman, "The Molecular Biology of *SRY* and Its Role in Sex Determination in Mammals," *Reproduction, Fertility, and Development* 7 (1995): 713–22.

4. The receptors for sex hormones are protein molecules located inside the cells. There is a receptor type for each type of sex hormone, and each matches its particular hormone in a "lock and key" fashion. Once the sex hormone has attached to its receptor, it has an effect on the genes in the cell's nucleus. It turns on certain genes, the end result being the formation of new proteins in the cell.

5. The control sequence is as follows: First, releasing factors are made by nerve cells in the hypothalamus and enter the portal bloodstream, which transports them to the anterior pituitary. The releasing factors then act on the cells of the anterior pituitary, causing them to release either luteinizing hormone or follicle-stimulating hormone into the general bloodstream, which carries them to the gonads. The hormones then stimulate the gonads to release the sex hormones into the blood, and these travel around the body. As the sex hormones also reach the hypothalamus, they provide feedback, indicating their own levels in the blood, and the hypothalamus adjusts its control accordingly to either raise or lower their levels via its effect on the pituitary. This feedback loop can also be regulated by inputs to the hypothalamus from higher centers in the brain (discussed further in chapter 3).

6. A classical publication that assumes the line of causation of sex differences to be from genes to hormones to brain function is R. W. Goy and B. S. McEwen, *Sexual Differentiation of the Brain* (Cambridge: MIT Press, 1980).

7. Frank Beach pointed this out nearly three decades ago: F. A. Beach, "Hormonal Factors Controlling the Differentiation, Development, and Display of Cop-

ulatory Behavior in the Ramstergig and Related Species," in E. Tobach, L. R. Aronson, and E. Shaw, eds., *The Biopsychology of Development* (New York: Academic Press, 1971).

8. J. Imperato-McGinley, R. E. Peterson, T. Gautier, and E. Sturia, "Androgens and the Evolution of Male Gender Identity Among Male Pseudohermaphrodites with 5–alpha-reductase Deficiency," *Acta Endocrinologica* 87 (1979): 259–69.

9. M. McGue and T. J. Bouchard, Jr., " Genetic and Environmental Influences on Human Behavioral Differences," *Annual Reviews in Neuroscience* 21 (1998): 1–24.

10. A. Feingolf, "Cognitive Gender Differences Are Disappearing," *American Psychologist* 43 (1988): 95–103.

11. J. S. Hyde, E. Fennema, and S. J. Lamon, "Gender Differences in Mathematics Performance: A Meta-analysis," *Psychology Bulletin* 107 (1990): 139–55.

12. M. S. Masters and B. Sanders, "Is the Gender Difference in Mental Rotation Disappearing?" *Behavior Genetics* 23 (1993): 337–42.

13. B. A. Gladue and J. M. Bailey, "Spatial Ability, Handedness, and Human Sexual Orientation," *Psychoneuroendocrinology* 20 (1995): 487–97; S. A. Berenbaum, K. Korman, and C. Leveroni, "Early Hormones and Sex Differences in Cognitive Abilities," *Learning and Individual Differences* 7 (1995): 303–21; D. J. Wegesin, "Relation Between Language Lateralization and Spatial Ability in Gay and Straight Women and Men," *Laterality* 3 (1998): 227–39.

14. D. Kimura, "Sex Differences in the Brain," *Scientific American* 267 (September 1992): 81–87 (quotation on 81).

15. Steven Rose has alerted scientists to the hierarchy of the levels, or orders, of analysis and the dangers of reductionism. See S. P. R. Rose, "From Causations to Translations: A Dialectical Solution to a Reductionist Enigma," in Rose, ed., *Towards a Liberatory Biology* (London: Allison and Busby, 1982); S. P. R. Rose, "Biological Reductionism: Its Roots and Social Functions," in L. Birke and J. Silvertown, eds., *More Than the Parts: Biology and Politics* (London: Pluto Press, 1985).

16. Sociobiology began with the publication of E. O. Wilson, *Sociobiology: A New Synthesis* (Cambridge: Harvard University Press, 1975). It has gained a large following, but perhaps the other most influential publication in the area, particularly for the nonspecialist, was R. Dawkins, *The Selfish Gene* (1976) (2d ed., New York: Oxford University Press, 1989).

17. See, for example, Wilson, *Sociobiology*.

18. M. Ridley, *The Red Queen: Sex and the Evolution of Human Nature* (New York: Penguin, 1994), 171–72.

19. G. Kaplan and L. J. Rogers, *The Orangutans* (New York: Perseus, 2000); also, St. Leonards (Australia): Allen and Unwin, 1999.

20. Some of the ideas of the evolutionary psychologists are outlined in R. Wright, *Time*, September 11, 1993. For a more comprehenesive account of evolutionary psychology, see S. Pinker, *How the Mind Works* (New York: Norton, 1997).

21. Summarized in C. N. Degler, *In Search of Human Nature: The Decline and*

Revival of Darwinism in American Social Thought (New York: Oxford University Press, 1991).

22. For elaboration on this topic see S. Rose, "The Rise of Neurogenetic Determinism," *Nature* 373 (1995): 380–83.

23. Pinker, *How the Mind Works*, 43.

24. E. O. Wilson, "Human Decency Is Animal," *New York Times Magazine*, October 12, 1975; cited in A. Fausto-Sterling, *Myths of Gender* (New York: Basic Books, 1985).

25. These particular notions have a long history and are covered in B. Lloyd and J. Archer, eds., *Exploring Sex Differences* (London: Academic Press, 1976).

26. These views can be found in a number of writings by sociobiologists and are stated plainly in D. Barash, *Sociobiology: The Whisperings Within* (London: Souvenir, 1980).

27. A. Moir and D. Jessel, *Brainsex: The Real Difference Between Men and Women* (reissue; New York: Dell Books, 1993).

28. This theory is summarized in D. F. Sherry and E. Hampson, "Evolution and the Hormonal Control of Sexually-Dimorphic Spatial Abilities in Humans," *Trends in Cognitive Sciences* 1 (1997): 50–56.

29. S. J. C. Gaulin and R. W. Fitzgerald, "Sex Differences in Spatial Ability: An Evolutionary Hypothesis and Test," *American Naturalist* 127 (1986): 74–88.

30. S. J. C. Gaulin and R. W. Fitzgerald, "Sex Differences in Spatial Ability and Activity in Two Vole Species (*Microtus ochrogaster* and *Microtus pennsylvanicus*)," *Journal of Comparative Psychology* 104 (1990): 88–93; L. F. Jacobs and W. D. Spencer, "Natural Space-use Patterns and Hippocampal Size in Kangaroo Rats," *Brain, Behavior, and Evolution* 44 (1994): 125–32.

31. M. Eals and I. Silverman, "The Hunter-Gatherer Theory of Spatial Sex Differences: Proximate Factors Mediating the Female Advantage in Recall of Object Arrays," *Ethology and Sociobiology* 15 (1994): 95–105.

32. D. H. McBurney, S. C. G. Gaulin, T. Devineni, and C. Adams, "Superior Spatial Memory of Women: Stronger Evidence for the Gathering Hypothesis," *Evolution and Human Behavior* 18 (1997): 165–74.

33. J. W. Berry, "Temne and Eskimo Perceptual Skills," *International Journal of Psychology* 1 (1966): 207–29; J. W. Berry, "Ecological and Social Factors in Spatial Perceptual Development," *Canadian Journal of Behavioral Science* 3 (1971): 324–36.

34. The theory of symmetry and sexual success is summarized in D. Concar, "Sex and the Symmetrical Body," *New Scientist* 146 (April 1995): 40–44.

35. R. Thornhill, S. W. Gangestad, and R. Comer, "Human Female Orgasm and Mate Fluctuating Asymmetry," *Animal Behavior* 50 (1995): 1601–15.

3. Gay Genes?

1. S. Rose, *Lifelines: Biology, Freedom, Determinism* (New York: Oxford University Press, 2000).

2. For further discussion on this point see S. Rose, "The Rise of Neurogenetic Determinism," *Nature* 373 (1995): 380–83.

3. J. Horgan, "Eugenics Revisited," *Scientific American* 268 (June 1993): 93–100.

4. The concept of statistical normality forms the basis of the following book, which has received much attention by the media: R. J. Herrnstein and C. A. Murray, *The Bell Curve: Intelligence and Class Structure in American Life* (New York: Free Press, 1994).

5. There are many papers on this topic but see G. Phillips and R. Over, "Differences Between Heterosexual, Bisexual, and Lesbian Women in Recalled Childhood Experiences," *Archives of Sexual Behavior* 24 (1995): 1–20; R. C. Pillard and J. M. Bailey, "Human Sexual Orientation Has a Heritable Component," *Human Biology* 70 (1998): 347–65.

6. The Kinsey report found that 70 percent of men in North America reported that they had reached orgasm in at least one homosexual encounter: A. C. Kinsey, W. B. Pomeroy, and C. E. Martin, *Sexual Behavior in the Human Male* (Philadelpia: Saunders, 1948). Other, more recent studies have reported similar figures.

7. D. H. Hamer, S. Hu, V. L. Magnuson, N. Hu, and A. M. L. Pattatucci, "A Linkage Between DNA Markers on the X Chromosome and Male Sexual Orientation," *Science* 261 (1993): 321–27; summarized in S. LeVay and D. H. Hamer, "Evidence for a Biological Influence in Male Homosexuality," *Scientific American* 269 (May 1994): 20–25.

8. D. H. Hamer, S. Hu, V. L. Magnuson, N. Hu, and A. M. L. Pattatucci, "A Linkage Between DNA Markers," 382.

9. R. C. Pillard and J. M. Bailey, "A Biological Perspective on Sexual Orientation," *The Psychiatric Clinics of North America* 18 (1995): 71–84.

10. For a critical discussion of the statistical methodology used by Hamer and colleagues, see N. Risch, E. Squires-Wheeler, and B. J. B. Keats, "Males Sexual Orientation and Genetic Evidence," *Science* 262 (1993): 2063–65, including a reply from Hamer, Hu, Magnuson, Hu, and Pattatucci ("A Linkage Between DNA Markers") on 2065. At this writing, the recent study failing to confirm Hamer's finding is G. Rice, C. Anderson, N. Risch, and G. Ebers, "Male Homosexuality: Absence of Linkage to Microsatellite Markers at Xq28," *Science* 284 (1999): 665–67.

11. B. Bower, "Genetic Clue to Male Homosexuality Emerges," *Science News*, July 17, 1993, 37.

12. Summarized in S. K. Miller, "Gene Hunters Sound Warning Over Gay Link," *New Scientist*, July 24, 1993, 4–5.

13. S. Hu et al., "Linkage Between Sexual Orientation and Chromosome *Xq28* in Males But Not Females," *Nature Genetics* 11 (1995): 248–56.

14. R. Pool, "Evidence for Homosexuality Gene," *Science* 261 (1993): 291–92.

15. For details of methodology see N. Martin, D. Boomsma, and G. Machin, "A Twin-pronged Attack on Complex Traits," *Nature Genetics* 17 (1997): 387–92.

16. J. M. Bailey and A. P. Bell, "Familiality of Female and Male Homosexual-

ity," *Behavior Genetics* 23 (1993): 313–22; A. M. L. Pattatucci and D. H. Hamer, "Development and Familiality of Sexual Orientation in Females," *Behavior Genetics* 25 (1995): 407–20.

17. The study of twins has been applied to many characteristics or traits. See P. Mittler, *The Study of Twins* (Harmondsworth: Penguin, 1971).

18. A. J. DeCasper and W. P. Fifer, "Of Human Bonding: Newborns Prefer Their Mothers' Voices," *Science* 208 (1980): 1174–76; A. J. DeCasper and A. D. Sigafoos, "The Intrauterine Heartbeat: A Potent Reinforcer for Newborns," *Infant Behavior and Development* 6 (1983): 19–25; W. P. Fifer, "Neonatal Preference for Mother's Voice," in N. A. Krasnegor, E. M. Blass, M. A. Hofer, and W. P. Smotherman, eds., *Perinatal Development: A Psychobiological Perspective*, 111–24 (Orlando, Fla.: Academic Press, 1987).

19. L. Kamin, *The Science and Politics of IQ* (Mahwah, N.J.: Lawrence Erlbaum Associates, 1974).

20. J. M. Bailey and R. C. Pillard, "A Genetic Study of Male Sexual Orientation," *Archives of General Psychiatry* 48 (1991): 1089–96. See also R. C. Pillard and J. M. Bailey, "A Biologic Perspective on Sexual Orientation," 71–84.

21. J. M. Bailey, R. C. Pillard, M. C. Neale, and Y. Agyei, "Heritable Factors Influence Sexual Orientation in Women," *Archives of General Psychiatry* 50 (1993): 217–23.

22. T. Lidz, "Reply to a Genetic Study of Male Sexual Orientation," *Archives of General Psychiatry* 50 (1993): 240.

23. E. D. Eckert, T. J. Bouchard, J. Bohlen, and L. L. Heston, "Homosexuality in Monozygotic Twins Reared Apart," *British Journal of Psychiatry* 14 (1986): 421–25.

24. This study, conducted by J. M. Bailey and colleagues, is cited in R. C. Pillard and J. M. Bailey, "Human Sexual Orientation Has a Heritable Component," 347–65.

25. S. LeVay and D. H. Hamer, "Evidence for a Biological Influence in Male Homosexuality," 20–25.

26. S. LeVay, "A Difference in Hypothalamic Structure Between Heterosexual and Homosexual Men," *Science* 253 (1991): 1034–37; S. LeVay, *The Sexual Brain* (Cambridge: MIT Press, 1993).

27. L. S. Allen and R. A. Gorski, "Sexual Orientation and the Size of the Anterior Commissure in the Human Brain," *Proceedings of the National Academy of Sciences USA* 89 (1992): 7199–7202.

28. M. Hines, "Hormonal and Neural Correlates of Sex-typed Behavioral Development in Human Beings," In M. Haug, R. E. Whalen, C. Aron, and K. L. Olsen, eds., *The Development of Sex Differences and Similarities in Behavior*, 131–49 (Norwell, Mass.: Kluwer, 1993).

29. D. F. Swaab and M. A. Hofman, "An Enlarged Suprachiasmatic Nucleus in Homosexual Men," *Brain Research* 587 (1990): 141–48.

30. D. F. Swaab and M. A. Hofman, "Sexual Differentiation of the Human Hy-

pothalamus in Relation to Gender and Sexual Orientation," *Trends in Neurosciences* 18 (1995): 264–70.

31. LeVay, *The Sexual Brain*, 1.

32. For further discussion of these issues see A. S. Greenberg and J. M. Bailey, "Do Biological Explanations of Homosexuality Have Moral, Legal, or Policy Implications?" *Journal of Sex Research* 30 (1993): 245–51.

33. J. P. Rushton, "Race Differences in Behavior: A Review and Evolutionary Analysis," *Personalities and Individual Differences* 9 (1988): 1009–24. For a critique of this work see E. Tobach and B. Rosoff, eds., *Challenging Racism and Sexism*, Genes and Gender 7 (New York: Femminist/City University of New York, 1994).

34. W. A. Henry III, "Born Gay?" *Time*, July 26, 1993, 44–47. See also J. Maddox, "Is Homosexuality Hard-wired?" *Nature* 353 (1991): 13.

35. G. Vines, "Obscure Origins of Desire," *New Scientist*, November 22, 1992, 2–8.

36. M. Barinaga, "Is Homosexuality Biological?" *Science* 253 (1991): 956–57.

37. For an example, see R. Dawkins, *The Selfish Gene* (1976) (2d ed., New York: Oxford University Press, 1989). Also see D. Hamer and P. Copeland, *Living with Our Genes* (New York: Doubleday, 1998).

38. R. Hubbard, "Race and Sex as Biological Categories," in Tobach and Rosoff, eds., *Challenging Racism and Sexism*, 11–21.

4. Hormones, Sex, and Gender

1. R. M. Rose, P. Bourne, and R. Poe, "Androgen Responses to Stress," *Psychosomatic Medicine* 31 (1969): 418–36.

2. A. Mazur and T. A. Lamb, "Testosterone, Status, and Mood in Human Males," *Hormones and Behavior* 14 (1980): 236–46; A. Booth, G. Shelley, A. Mazur, G. Tharp, and R. Kittok, "Testosterone and Winning and Losing Human Competition," *Hormones and Behavior* 23 (1989): 556–71; K. McCaul, B. Gladue, and M. Joppa, "Winning, Losing, Mood, and Testosterone" *Hormones and Behavior* 26 (1992): 486–506.

3. For an introduction see S. M. Breedlove, "Sexual Differentiation of the Brain and Behavior," in J. B. Becker, S. M. Breedlove, and D. Crews, eds., *Behavioral Endocrinology*, 39–68 (Cambridge: MIT Press, 1992).

4. K. Larsson, "Experiential Factors in the Development of Sexual Behaviour," in J. B. Hutchinson, ed., *Biological Determinants of Sexual Behavior* (Chichester, Eng.: Wiley, 1978).

5. J. D. Blaustein and D. H. Osler, "Gonadal Steroid Hormone Receptors and Social Behaviors," in J. Balthazart, ed., *Advances in Comparative and Environmental Physiology 3: Molecular and Cellular Basis of Social Behavior in Vertebrates*, 31–104 (Berlin: Springer, 1989).

6. R. W. Goy and D. A. Goldfoot, "Experiential and Hormonal Factors Influencing Development of Sexual Behavior in the Male Rhesus Monkey," in F. O.

Schmitt and F. G. Worden, eds., *The Neurosciences: Third Study Program*, 571–81 (Cambridge: MIT Press, 1974). For further information on this topic see I. L. Ward, "Sexual Behavior: The Product of Perinatal Hormonal and Prepuberial Social Factors," in A. A. Gerall, H. Moltz, and I. L. Ward, eds., *Sexual Differentiation*, 157–80. Volume 11 of *Handbook of Behavioral Neurobiology* (New York: Plenum, 1992).

7. C. S. Carter, "Hormonal Influences on Human Sexual Behavior," in Becker, Breedlove, and Crews, eds., *Behavioral Endocrinology*, 131–42.

8. G. I. M. Swyer, "Clinical Effects of Agents Affecting Fertility," in R. P. Michael, ed., *Endocrinology and Human Behavior* (Oxford: Oxford University Press, 1968).

9. C. Fabre-Nys, "Steroid Control of Monoamines in Relation to Sexual Behavior," *Reviews of Reproduction* 3 (1998): 31–41.

10. K. Wallen, "Desire and Ability: Hormones and the Regulation of Female Sexual Behavior," *Neuroscience and Biobehavioral Reviews* 14 (1990): 233–41.

11. B. B. Sherwin, "A Comparative Analysis of the Role of Androgen in Human Male and Female Sexual Behavior: Behavioral Specificity, Critical Thresholds, and Sensitivity," *Psychobiology* 16 (1988): 416–25.

12. D. W. Pfaff, *Estrogens and Brain Function* (Berlin: Springer, 1980). See also D. W. Pfaff and M. Keiner, "Atlas of Estradiol-Concentrating Cells in the Central Nervous System of the Female Rat," *Journal of Comparative Neurology* 151 (1973): 121–58.

13. J. Balthazart, "Steroid Metabolism and the Activation of Social Behavior," in Balthazart, ed., *Advances in Comparative and Environmental Physiology 3*, 105–109.

14. J. Balthazart and G. F. Ball, "New Insights into Regulation and Function of Brain Estrogen Synthetase (aromatase)," *Trends in Neurosciences* 21 (1998): 243–49. For important earlier work on receptors for steroid hormones in the brain, see papers by Bruce McEwen: for example, B. S. McEwen et al., "Steroid Hormone Receptors," in K. Fuxe, T. Hïkfelt, and R. Luft, eds., *Central Regulation of the Endocrine System*, 261–71 (New York: Plenum, 1979); B. S. McEwen, I. Lieberburg, C. Chaptal, and L. C. Krey, "Aromatization: Important for Sexual Differentiation of the Neonatal Rat Brain," *Hormones and Behavior* 9 (1977): 249–63.

15. D. M. Broverman, I. K. Broverman, W. Vogel, and R. D. Palmer, "The Automatization Cognitive Style and Physical Development," *Child Development* 35 (1964): 1343–59; E. L. Klaiber, D. M. Broverman, and Y. Kobayashi, "The Automatization Cognitive Style, Androgens, and Monoamine Oxidase," *Psychopharmacology* 11 (1967): 320–36: summarized in L. J. Rogers, "Male Hormones and Behavior," in B. Lloyd and J. Archer, eds., *Exploring Sex Differences*, 157–84 (London: Academic Press, 1976).

16. D. M. Broverman, E. L. Klaiber, and W. Vogel, "Gonadal Hormones and Cognitive Functioning," in J. E. Parsons, ed., *The Psychobiology of Sex Differences and Sex Roles*, 57–80 (New York: McGraw-Hill, 1980).

17. D. M. Broverman, "Generality and Behavioral Correlates of Cognitive Styles," *Journal of Consulting Psychology* 28 (1964): 487–500.

18. P. Caplan, G. MacPherson, and P. Tobin, "Do Sex-related Differences in Spatial Abilities Exist?" *American Psychologist* 40 (1985): 786–99.

19. E. Hampson and D. Kimura, "Sex Differences and Hormonal Influences on Cognitive Function in Humans," in Becker, Breedlove, and Crews, eds., *Behavioral Endocrinology*, 357–400.

20. V. J. Shute, J. W. Pelligrino, L. Hurbert, and R. W. Reynolds, "The Relationship Between Androgen Levels and Human Spatial Abilities," *Bulletin of the Psychonomic Society* 21 (1983): 465–68.

21. W. F. McKeever, "Hormone and Hemisphericity Hypothesis Regarding Cognitive Sex Differences: Possible Future Explanatory Power, But Current Empirical Chaos," *Learning and Individual Differences* 7 (1995): 323–40.

22. G. W. Harris and S. Levine, "Sexual Differentiation of the Brain and Its Experimental Control," *Journal of Physiology* 181 (1965): 379–400.

23. C. H. Phoenix, R. W. Goy, and J. A. Resko, "Psychosexual Differentiation as a Function of Androgenic Stimulation," in H. Diamond, ed., *Perspectives in Reproduction and Sexual Behavior* (Bloomington: Indiana University Press, 1968); C. H. Phoenix, "Prenatal Testosterone in the Nonhuman Primate and Its Consequences for Behavior," in R. C. Friedman, R. M. Richart, and R. L. Vandeeds, eds., *Sex Differences in Behavior*, 19–32 (New York: Wiley, 1974).

24. R. W. Goy, W. E. Bridson, and W. C. Young, "Period of Maximal Susceptibility to the Prenatal Female Guinea Pig to Masculinizing Actions of Testosterone Proprionate," *Journal of Comparative and Physiological Psychology* 57 (1964): 166–74.

25. J. Money and A. A. Erhdardt, *Man and Woman, Boy and Girl: The Differentiation and Dimorphism of Gender Identity from Conception to Maturity* (Baltimore: Johns Hopkins University Press, 1972); J. Money and A. A. Erhdardt, "Progestin-induced Hermaphroditism: IQ and Psychosexual Identity in the Study of Ten Girls," *Journal of Sex Research* 3 (1967): 83–100.

26. S. W. Baker and A. A. Ehrhardt, "Pre-natal Androgen, Intelligence, and Cognitive Sex Differences," in Friedman, Richart, and Vandeeds, eds., *Sex Differences in Behavior*, 53–84.

27. S. Berenbaum and M. Hines, "Early Androgens Are Related to Childhood Sex-typed Toy Preference," *Psychological Science* 3 (1992): 203–206; S. Berenbaum and E. Snyder, "Early Hormonal Influences on Childhood Sex-typed Activity and Playmate Preference: Implications for the Development of Sexual Orientation," *Developmental Psychology* 31 (1995): 31–42.

28. L. J. Rogers and J. Walsh, "Short-comings of Psychomedical Research into Sex Differences in Behavior: Social and Political Implications," *Sex Roles* 8 (1982): 269–81; L. J. Rogers, "Hormonal Theories for Sex Differences: Politics Disguised as Science," *Sex Roles* 9 (1983): 1109–13.

29. F. Slijper, "Androgens and Gender Role Behavior in Girls with Congenital Andrenal Hyperplasia," in G. J. De Vries, J. P. C. DeBruin, H. M. B. Uylings, and M. A. Corner, eds., *Progress in Brain Research*, 417–22 (Amsterdam: Elsevier, 1984).

30. R. W. Dittmann, M. E. Kappes, and M. H. Kappes, "Sexual Behavior in Adolescent and Adult Females with Congenital Adrenal Hyperplasia," *Psychoneuroendrocrinology* 17 (1992): 153–70.

31. J. Money and J. Dalery, "Iatrogenic Homosexuality," *Journal of Homosexuality* 1 (1976): 357–71; G. Dörner, "Hormones and Sexual Differentiation of the Brain," *Sex, Hormones, and Behavior* (Ciba Foundation Symposium) 62 (1979): 81–112; G. Dörner, *Hormones and Brain Differentiation* (Amsterdam: Elsevier, 1976); J. M. Reinisch, M. Ziemba-Davis, and S. A. Sanders, "Hormonal Contributions to Sexually Dimorphic Behavioral Development in Humans," *Psychoneuroendocrinology* 16 (1991): 213–78. For a critical discussion of this claim, see L. Birke, "Is Homosexuality Hormonally Determined?" *Journal of Homosexuality* 6 (1979): 35–50.

32. I. L. Ward and J. Weisz, "Maternal Stress Alters Plasma Testosterone in Fetal Males," *Science* 207 (1980): 328–29.

33. G. Dörner et al., "Prenatal Stress as Possible Etiogenetic Factor of Homosexuality in Human Males," *Endokrinologie* 75 (1980): 365–86; G. Dörner, B. Schenk, B. Schmiedel, and L. Ahrens, "Stressful Events in Prenatal Life of Bi- and Homosexual Men," *Experimental and Clinical Endocrinology* 81 (1983): 83–87.

34. N. Geschwind and A. M. Galaburda, *Cerebral Lateralization: Biological Mechanisms, Associations, and Pathology* (Cambridge: MIT Press, 1987).

35. G. Vines, "Obscure Origins of Desire," *New Scientist*, November 22, 1992, 2–8.

36. Reviewed by W. Byne and B. Parsons, "Human Sexual Orientation," *Archives of General Psychiatry* 50 (1993): 228–39.

37. C. M. McCormick, S. F. Witelson, and E. Kingstone, "Left-handedness in Homosexual Men and Women: Neuroendocrine Implications," *Psychoneuroendocrinology* 15 (1990): 69–76.

38. R. Green, "Dimensions of Human Sexual Identity: Transsexuals, Homosexuals, Fetishists, Cross-gendered Children, and Animal Models," in M. Haug, R. E. Whalen, C. Aron, and K. L. Olsen, eds., *The Development of Sex Differences and Similarities in Behavior*, 477–86 (Norwell, Mass.: Kluwer, 1993); Dörner, *Hormones and Brain Differentiation*.

39. J. A. Will, P. A. Self, and N. Datan, "Maternal Behavior and Perceived Sex of Infant," *American Journal of Orthopsychiatry* 46 (1976): 135–40.

40. J. Imperato-McGinley, R. E. Peterson, T. Gautier, and E. Sturla, "Androgens and the Evolution of Male Gender Identity Among Male Psuedohermaphrodites with 5–alpha-reductase Deficiency," *Acta Endocrinologica* 87 (1979): 259–69.

41. J. Money, "Ablatio Penis: Normal Male Infant Sex Reassignment as a Girl,"

Archives of Sexual Behavior 4 (1975): 65–71. See also J. Colapinto, *As Nature Made Him: The Boy Who Was Raised as a Girl* (New York: HarperCollins, 2000).

42. M. Diamond and H. K. Sigmundson, "Sex Reassignment at Birth: Long-term Review and Clinical Implications," *Archives of Pediatric and Adolescent Medicine* 151 (1997): 298–304; also Colapinto, *As Nature Made Him.*

43. C. L. Moore and G. A. Moreli, "Mother Rats Interact Differently with Male and Female Offspring," *Journal of Comparative and Physiological Psychology* 93 (1979): 677–84.

44. M. J. Baum, S. C. Bressler, M. C. Daum, C. A. Veiga, and C. S. McNamee, "Ferret Mothers Provide More Anogenital Licking to Male Offspring: Possible Contribution to Psychosexual Differentiation," *Physiology and Behavior* 60 (1996): 353–59.

45. C. L. Moore, "Maternal Contributions to Mammalian Reproductive Development and the Divergences of Males and Females," *Advances in the Study of Behavior* 24 (1995): 47–118; C. L. Moore, "Maternal Contributions to the Development of Masculine Sexual Behavior in Laboratory Rats," *Developmental Psychobiology* 17 (1984): 347–56.

46. C. L. Moore, "Maternal Behavior of Rats Is Affected by Hormonal Condition of Pups," *Journal of Comparative and Physiological Psychology* 1 (1982): 123–29.

47. C. L. Moore, "An Olfactory Basis for Maternal Discrimination of Sex of Offspring in Rats," *Animal Behavior* 29 (1981): 383–86; C. L. Moore, "Sex Differences in Urinary Odors Produced by Young Laboratory Rats," *Journal of Comparative Psychology* 99 (1985): 336–41.

48. C. L. Moore, H. Dou, and J. M. Juraska, "Maternal Stimulation Affects the Number of Motor Neurons in a Sexually Dimorphic Nucleus of the Lumbar Spinal Cord," *Brain Research* 572 (1992): 52–56.

49. C. L. Moore, "Another Psychobiological View of Sexual Differentiation," *Development Reviews* 5 (1985): 18–55.

50. G. Vines, *Raging Hormones: Do They Rule Our Lives?* (Berkeley: University of California Press, 1994).

51. J. Archer, "The Influence of Testosterone on Human Aggression," *British Journal of Psychology* 82 (1991): 1–28.

5. Experience, Interactions, and Change

1. More information on the corpus callosum and its role in hemispheric dominance can be found in M. C. Corballis, *The Lopsided Ape: Evolution of the Generative Mind* (New York: Oxford University Press, 1991); N. Geschwind and A. M. Galaburda, *Cerebral Dominance: The Biological Foundations* (Cambridge: Harvard University Press, 1984); J. B. Hellige, *Hemispheric Asymmetry: What's Right and What's Left* (Cambridge: Harvard University Press, 1993).

2. Some of the historical aspects of claimed sex differences in lateralization are discussed in S. L. Star, "The Politics of Right and Left: Sex Differences in Hemi-

spheric Brain Asymmetry," in R. Hubbard, M. S. Henifin, and B. Fried, eds., *Women Look at Biology Looking at Women*, 51–74 (Cambridge: Schenkman, 1979). For the hypothesis that women's brains are less lateralized than men's, see J. Levy, "The Mammalian Brain and the Adaptive Advantage of Cerebral Asymmetry," *Annals of the New York Academy of Science* 229 (1977): 265–72. For the contrary hypothesis, stating that women's brains are less lateralized than men's, see A. W. H. Buffery, "Male and Female Brain Structure and Function: Neuropsychological Analyses," in N. Grieve and P. Grimshaw, eds., *Australian Women: New Feminist Perspectives* (New York: Oxford University Press, 1986). See also J. McGlone, "Sex Difference in Human Brain Asymmetry: A Critical Survey," *Behavioral and Brain Sciences* 3 (1980): 215–63.

3. M. Kinsbourne, "If Sex Differences in the Brain Exist, They Have Yet to Be Discovered," *Behavioral and Brain Sciences* 3 (1980): 241–42.

4. The reviews of sex differences in lateralization in the human brain are summarized in Hellige, *Hemispheric Asymmetry*, 232–39.

5. C. de Lacoste-Utamsing and R. L. Holloway, "Sexual Dimorphism in the Human Corpus Callosum," *Science* 216 (1982): 1431–32; M. C. de Lacoste, R. L. Holloway, and D. J. Woodward, "Sex Differences in the Fetal Corpus Callosum," *Human Neurobiology* 5 (1986): 93–96; L. S. Allen, M. F. Richey, Y. M. Chai, and R. A. Gorski, "Sex Differences in the Corpus Callosum of the Living Human Being," *Journal of Neuroscience* 11 (1991): 933–42.

6. A. Murr and A. Rogers, "Gray Matters," *The Bulletin*, March 28, 1995, 76–82.

7. S. F. Witelson, "The Brain Connection: The Corpus Callosum Is Larger in Left-handers," *Science* 229 (1985): 665–68; S. F. Witelson, "Hand and Sex Differences in the Isthmus and Genu of the Human Corpus Callosum," *Brain* 112 (1989): 799–835.

8. For example, see F. Aboitiz, A. B. Scheibel, and E. Zaidel, "Morphometry of the Sylvian Fissure and the Corpus Callosum, with Emphasis on Sex Differences," *Brain* 115 (1992): 1521–41.

9. L. E. Emory, D. H. Williams, C. M. Cole, E. G. Amparo, and W. J. Meyer, "Anatomic Variation of the Corpus Callosum in Persons with Gender Dysophoria," *Archives of Sexual Behavior* 20 (1991): 409–17. Also summarized in Hellige, *Hemispheric Asymmetry*, 128–29. See also A. D. Bell and S. Variend, "Failure to Demonstrate Sexual Dimorphism of the Corpus Callosum in Childhood," *Journal of Anatomy* 143 (1985): 143–47.

10. K. M. Bishop and D. Wahlsten, "Sex Differences in the Human Corpus Callosum: Myth or Reality?" *Neuroscience and Biobehavioral Reviews* 21 (1997): 581–601 (quotation from 593).

11. R. H. Fitch and V. H. Denenberg, "A Role for Ovarian Hormones in Sexual Differentiation of the Brain," *Behavioral and Brain Sciences* 21 (1998): 311–52.

12. V. H. Denenberg, "Hemispheric Laterality in Animals and the Effects of

Early Experience," *Behavioral and Brain Sciences* 4 (1981): 1–49; summarized in J. L. Bradshaw and L. J. Rogers, *The Evolution of Lateral Asymmetries, Language, Tool Use, and Intellect* (San Diego: Academic Press, 1993).

13. A. S. Berrebi et al., "Corpus Callosum: Region-specific Effects of Sex, Early Experience, and Age," *Brain Research* 438 (1988): 216–24. This sex difference in size was later shown to result, at least in part, from more of the nerve cell, axons in the corpus callosum being myelinated in the male than in the female. Myelinated nerve cell axons transmit information faster because the coating of myelin around the axon aids electrical conduction. This finding is reported in C. M. Mack, G. W. Boehm, A. S. Berrebi, and V. H. Denenberg, "Sex Differences in the Distribution of Axon Types Within the Genu of the Rat Corpus Callosum," *Brain Research* 697 (1995): 152–56.

14. V. H. Denenberg, R. H. Fitch, L. M. Schrott, P. E. Cowell, and N. S. Waters, "Corpus Callosum: Interactive Effects of Infantile Handling and Testosterone in the Rat," *Behavioral Neuroscience* 105 (1991): 562–66.

15. W. P. Smotherman, C. P. Brown, and S. Levine, "Maternal Responsiveness Following Differential Pup Treatment and Mother-Pup Interactions," *Hormones and Behavior* 8 (1977): 242–53.

16. P. E. Cowell, L. S. Allen, N. S. Zalatimo, and V. H. Denenberg, "A Developmental Study of Sex and Age Interactions in the Human Corpus Callosum," *Developmental Brain Research* 66 (1992): 187–92.

17. R. J. Andrew and A. Brennan, "Sex Differences in Lateralization in the Domestic Chick," *Neuropsychologia* 22 (1984): 503–509.

18. J. V. Zappia and L. J. Rogers, "Sex Differences and the Reversal of Brain Asymmetry by Testosterone in Chickens," *Behavioural Brain Research* 23 (1987): 261–67; L. J. Rogers, "Behavioral, Structural, and Neurochemical Asymmetries in the Avian Brain: A Model System for Studying Visual Development and Processing," *Neuroscience and Biobehavioral Reviews* 20 (1996): 487–503. Also summarized in L. J. Rogers, *The Development of Brain and Behavior in the Chicken* (New York: CAB International, 1995).

19. L. J. Rogers, "Light Experience and Asymmetry of Brain Function in Chickens," *Nature* 297 (1982): 223–25; L. J. Rogers, "Light Input and the Reversal of Functional Lateralization in the Chicken Brain," *Behavioural Brain Research* 38 (1990): 211–21.

20. L. J. Rogers and H. S. Sink, "Transient Asymmetry in the Projections of the Rostral Thalamus to the Visual Hyperstriatum of the Chicken, and Reversal of Its Direction by Light Exposure," *Experimental Brain Research* 70 (1988): 378–84.

21. S. Rajendra and L. J. Rogers, "Asymmetry Is Present in the Thalamofugal Visual Projections of Female Chicks," *Experimental Brain Research* 92 (1993): 542–44.

22. I. M. Schwartz and L. J. Rogers, "Testosterone: A Role in the Development of Brain Asymmetry in the Chick," *Neuroscience Letters* 146 (1992): 167–70; L. J.

Rogers and S. Rajendra, "Modulation of the Development of Light-initiated Asymmetry in Chick Thalamofugal Visual Projections by Estradiol," *Experimental Brain Research* 93 (1993): 89–94.

23. P. Bateson, ed., *The Development and Integration of Behavior* (New York: Cambridge University Press, 1991).

24. S. Oyama, *The Ontogeny of Information: Developmental Systems and Evolution* (rev. ed., New York: Cambridge University Press, 2000).

25. M. B. Casey, M. M. Brabeck, and R. L. Nuttall, "As the Twig Is Bent: The Biology and Socialization of Gender Roles in Women," *Brain and Cognition* 27 (1995): 237–46.

26. J. M. Reinisch, M. Ziemba-Davis, and S. A. Sanders, "Hormonal Contributions to Sexually Dimorphic Behavioral Development in Humans," *Psychoneuroendocrinology* 16 (1991): 213–78.

27. S. Rose, *Lifelines: Biology, Freedom, Determinism* (New York: Oxford University Press, 2000).

Index